AMERICAN ODYSSEY

by INGVARD HENRY EIDE

AMERICAN ODYSSEY
The Journey of Lewis and Clark

Rand McNally & Company   Chicago • New York • San Francisco

Copyright © 1969 by Rand McNally & Company
Copyright 1969 under International Copyright Union
by Rand McNally & Company
All rights reserved
Library of Congress Catalog Card Number 71-77805
Printed in the United States of America
by Rand McNally & Company

First paperback printing, 1979

To Edith

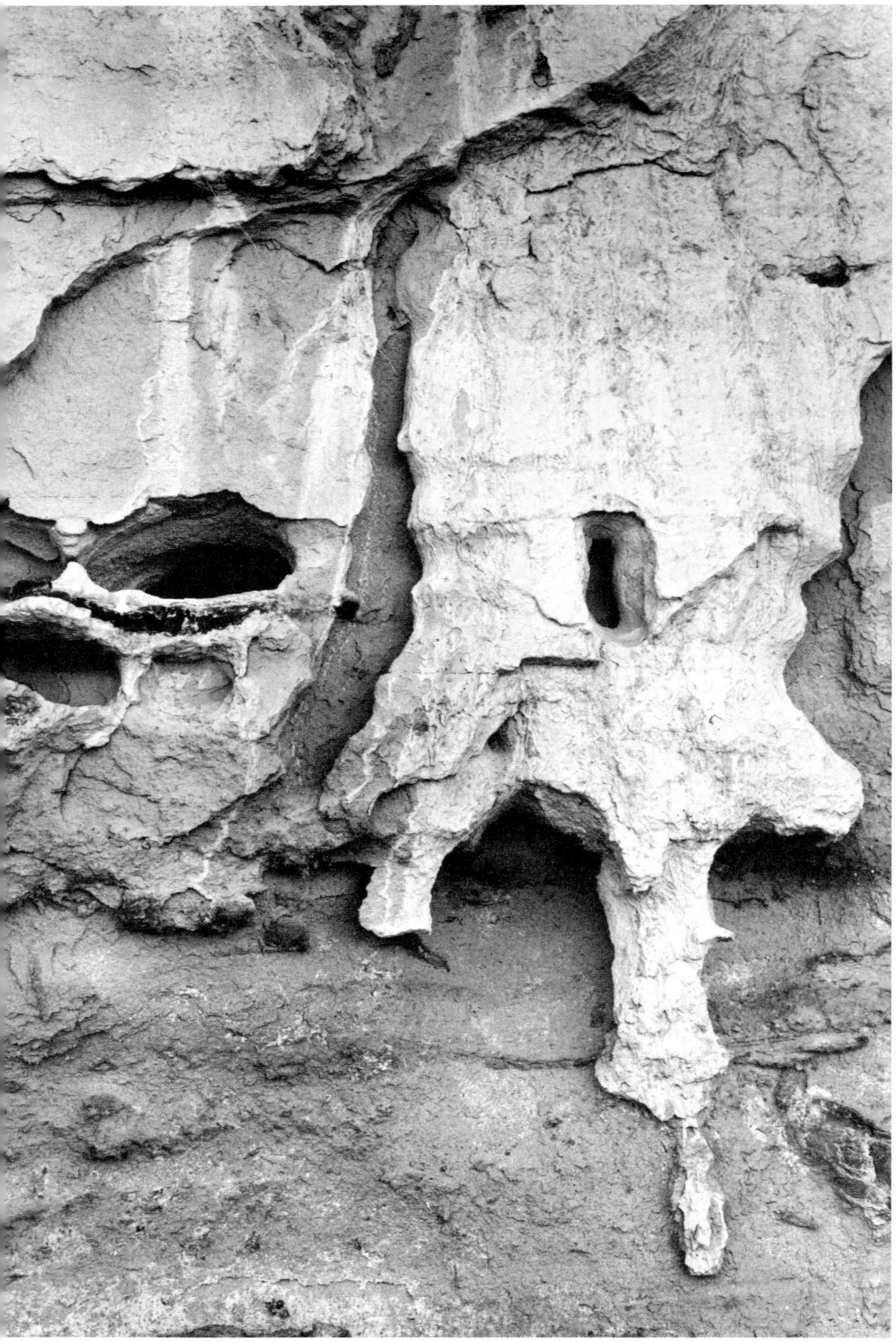
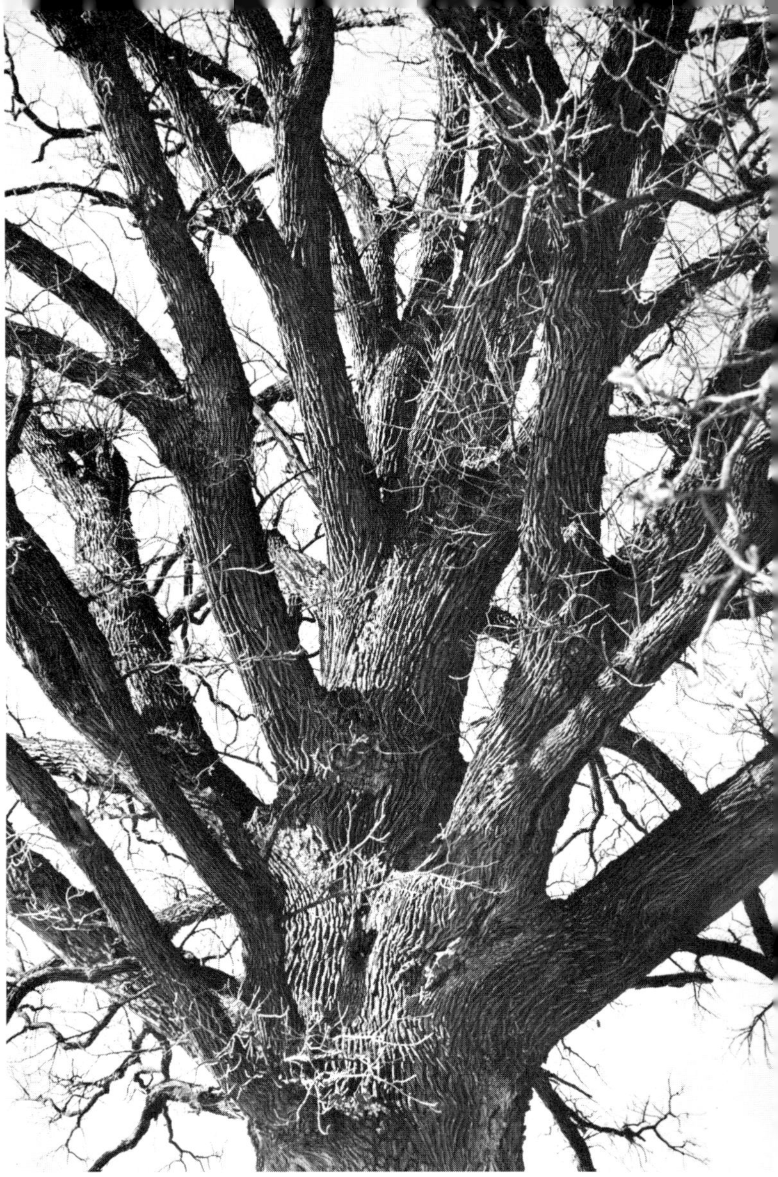
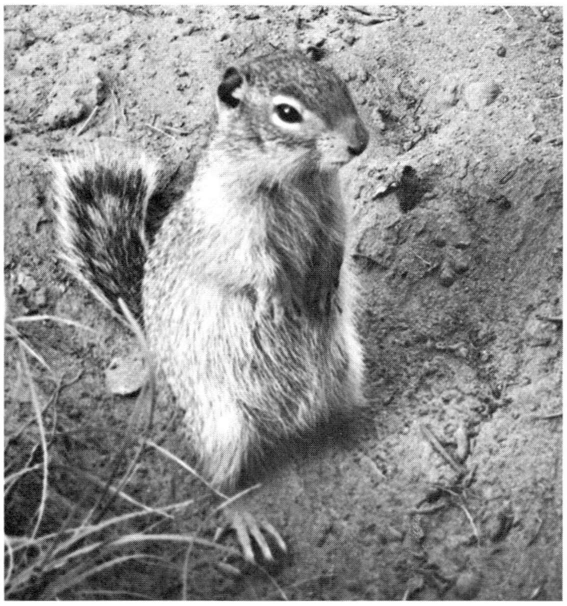

CONTENTS

Preface	ix
Acknowledgments	xv
Introduction	xix
The Plan	1
The Journey	21
List of Illustrations	231
A Note on the Photography	241
Sources	243
Books for Further Reading	245

Meriwether Lewis

William Clark

PREFACE

I stood on a high ridge of the Idaho timberlands an hour's drive from the Montana border. Around and below me lay a forest, the same forest and very likely some of the same trees that Meriwether Lewis and William Clark had seen more than a century and a half ago.

Lewis and Clark and thirty-one explorers had been there in the fall of 1805 on their way to the Pacific Ocean. The expedition left its winter headquarters at Wood River, Illinois, in 1804. Moving westward they crossed the Continental Divide and by the fall of 1805 they had reached this ridge in Idaho. They were close to starvation—not for the first time, nor the last. No game tracks showed in the snow.

The barrier of these forests and the Bitterroot Mountains was all the more vexing because the expedition was nearing its goal, the mouth of the Columbia River. But their courage and their devotion to purpose overcame the final barrier, and in November, 1805, Lewis and Clark and their men reached the Pacific.

The story of this overland march is the basis for my photographic chronicle. I have used their words and my camera. There is no historian's narrative. The original journals have an immediacy I could not equal. Passages from the diaries of the other Expedition members are used, in addition to the journals made by their leaders. The selections give more than the story of this undertaking. They tell, as nothing else could, of bravery, adventure, observation, and hardship which characterized it. Only fragments from the journals could be used in this book, but in their entirety the journals are a fascinating day-to-day account of a heroic American odyssey.

Their journey became my journey. The vast land, their new land, opened itself to my eyes as it did to the eyes of the Expedition. Their wonder was mine, their vision mine. The country has huge areas of wilderness, of willows and cottonwoods lining the Missouri River, of badlands with sparse, scrubby juniper, inhabited by prairie dogs and antelopes. It is a country with alkaline water, sweet water, wild water, of dense forests, deep ravines, canyons with cool alpine air, and of the great, muscled Columbia River, neighbor of three great peaks, Hood, Adams, and St. Helens.

It is almost too late for a historical record, yet some of the areas are still untouched. The Clearwater and Snake rivers, for example, and the mountain slopes and creeks of western Montana and Idaho. The magnificent Missouri River Breaks remain the same. Civilization has hardly touched Lewis and Clark Pass, Lemhi Pass, Lost Trail Pass, and the Two Medicine campsite where Lewis experienced the Expedition's only hostile encounter. However, long stretches of the Missouri in North and South Dakota, once rich with Expedition campsites, are forever lost. Other changes have occurred in the Midwest, where freeways override our heritage.

Even though the authentic Lewis and Clark view was not always to be found, my aim was to match my photographs as closely as possible with journal descriptions. In order to show details of topography near the campsites it was necessary in certain instances to take photographs out of season. Foliage, in these cases, would have obscured the scene itself. With these exceptions, each photograph was taken at the same time of year as the corresponding journal entry.

The journals reveal both happy and hazardous experiences. There are moments of glory, times of hardship and humor, and periods of uncertainty and not-quite despair. But always, on good day or ill, observation. This passion for information marks the men and the Expedition.

The scope of the journals is astonishing. Lewis and Clark were constantly aware of their responsibility to observe and record. Jefferson's objectives, wide in extent, were clearly stated: The Expedition was to explore the land, serve as a diplomatic mission, collect scientific information, and break trail for commerce. An earlier association is revealed, however, in Lewis's notation at Fort Clatsop in 1806; the Expedition also was for "those who may hereafter attemt the fir trade to the East Indies by way of the Columbia river and the Pacific Ocean"—the old dream of a transcontinental water route, a Northwest Passage. A transcontinental exploration of our country's western region occupied Jefferson even in 1783 when he asked George Rogers Clark, older brother of William, to "lead such a party." Nothing came of this, or of several other attempts, until two decades later when Jefferson's dream was made real by Meriwether Lewis and William Clark.

The original documents and letters printed in the beginning of this book establish the historical setting of the exploration. The journals themselves show how closely these men kept faith with their President and performed with high purpose the mission entrusted to them.

Here was remarkable dedication. It survived all difficulties, despite daily grind, cold, hunger, and disappointment.

To a photographer, a sequence of pictures of the journey was a challenge. And as these men were faithful to their undertaking, so have I tried to be faithful to them. My interest increased as I found more and more evidences of the Expedition's accomplishments. There could not be a mere scramble of views. There had to be chronology, of course, but also an unrolling of the land, a sense of discovery. I wanted also to reflect my own curiosity, shaped and enlarged by those early travelers, about a river, a bluff, or whatever remained of old beauty.

Photography as we know it today arrived half a century after the Expedition, but Lewis, sitting on the banks of the Great Falls of the Missouri, noted in his diary, Thursday, June 13, 1805, ". . . I most sincerely regreted that I had not brought a crimee [camera] obscura with me . . . ." as a means of illustrating the beauty he saw. He felt inadequate with words.

It is too bad that the personal incidents recorded in the journals antedated modern techniques of picture making. Photography could have shown a cast of nine young Kentuckians, the rowdy nucleus of the Expedition, near a tree on the right embankment of Silver Creek, Indiana, opposite Louisville. There, on the level terrain, was where William Clark drilled his nine volunteers prior to the arrival of Captain Lewis, in October of 1803.

Photography could have shown Private John Newman tried by the first court-martial in South Dakota, found guilty and sentenced to seventy-five blows on his bare back. Not even a pen-and-ink artist was on hand to record the expressions of the Indians who pleaded for Private Newman. Flogging, to the Indians, was incomprehensible, a barbarism.

Or think of the opportunities for photography when the Expedition lived with the Mandan Indians of North Dakota during the winter of 1804-05; or for the photographic descriptions of "gangues" of buffalo numbering 10,000, or of the mountain sheep, antelope, grizzly bears, deer, elk, and other animals. Or the reunion of Sacajawea and her brother, the Shoshone chief Cameahwait, on the east slope of the Bitterroot range; or of the exhausting labor at the portage at the Great Falls of the Missouri and later at the Great Falls of the Columbia.

Simple and direct photographic statements could have shown the happy eyes of those men as they saw wonders ahead, as they finally caught sight of the Pacific Ocean. Such a photographic story, of course, can never be shown; but I was fortunate, and, I hope, not too late to describe how the new country appeared to Lewis and Clark.

I considered the journals a first-rate script. Bernard DeVoto said they were an American classic, and, as he did in his interpretive study of the original journals, I found it necessary to edit my essay. Day by day I developed a format as I traveled from one site to another. Between journals and seen subjects I reached my decisions.

My editor, who was much in love with the journals and the country which they described, laid down no special requirements, except that the photographs should have some poetic freedom without sacrifice of truth.

Before the actual photographs were taken I experimented with locations and the uses of films, exposures, and filters. A habit had to be established that would simplify photography over the entire route. There were duplications of scenes, or near-duplications, and comparisons. Finally about one out of every dozen films was selected for publication.

To make the photographs more purposeful I chose to work under the same or almost-same weather conditions described in the reports. Every element was used, the threatening skies, clear days, windy weather, early morning and late evening hours. Twice in the same year I went back to the Pacific Coast to catch those gray and stormy times which were so frustrating to the explorers. It was not at all uncommon to go here and there the second or third time to improve on the first, or to make additional photographs when other subjects could be shot, and make certain that conditions were in accord with the journal descriptions.

The exclusive use of black-and-white films for this essay came not from the belief in its general superiority to color; it just seemed to suit my purpose better. With black-and-white I could catch more of mood, symbol, and the suggestion of space and loneliness. These were the moods so often expressed in the journals.

In my 57,000 mile, two-year retracing of the route of Lewis and Clark, I started at Silver Creek, near Clarksville, Indiana, across the river from Louisville, Kentucky. Silver Creek still flows, a little below the Falls of the Ohio, into the Ohio River. Here, too, is where Lewis and Clark joined forces.

From this point westward my ideas took shape. It became important to "see" the campsite, the embarkation point, the unusual. The scenes had to be compatible with the word descriptions and moods expressed in the journals. I had to capture and hold the feeling of wilderness.

Opposite Wood River, a jump or two above St. Louis, the Missouri pushes into the Mississippi. Unlike another frontier waterway, the respectable Ohio, from this point north and westward the Missouri becomes a rambunctious force with fast bends and winding reaches. Now the river became the Expedition's roadway.

Everywhere I noted change; these changes sometimes made verification of locations in the journals very difficult. For example, in the early days the Missouri came almost to the main street of St. Charles, the Expedition's second setting-off place. Clark described St. Charles as a picturesque town. It still is, but the Missouri has left the original shores and the scene is not the same. But with his words and those of Lewis, there is a glimpse of this old settlement.

So it is in so many places. The vigor of the river and the strength of its spring torrents have made a new shoreline, have traced a new bed. The river has torn and heaved at its banks, and in many places has sought, it almost seems, to widen the gap between what is civilized and what is not.

Farther west, into the rolling hills, prairies, mountains, and canyons, the actual locations along the route are a little easier to match with the journals. Many, many times I stood on almost the exact spot from which the Lewis and Clark party marvelled at the sight of their new territory. I do not regret the omission of certain sites and areas, some of which were deliberate. No staging of scenes was required.

A crowning memory of my travels on the Lewis and Clark trail was seeing the pleasure of local historians when I inquired about the Expedition for my pictorial story. Everywhere I went they increased my interest in the Expedition and in the amount of historical research being done, and my perspective opened to greater new vistas. I shall treasure their eagerness to assist me. Among those who have become near-expert on the movements of the Expedition I shall remember C. C. Gale, the eighty-year-old operator of a small fixit shop in out-of-the-way Rulo, Nebraska, and Ursula Higgins, Indian housewife and Community Action leader in Browning, Montana. She made me proud to know her people, the Blackfoot Indians. Many other people along the winding, sometimes hectic, route of Lewis and Clark have provided me with unforgettable memories and associations. Scientists, bankers, mechanics, schoolteachers, young people—all showed enthusiastic concern for my visual recapturing of the Expedition's journey. And of the journey's almost endless vistas, unforgettable to me are the unique Missouri River Breaks, the beauty of Packer Meadow and of Pack Creek, the solemnity of the Lochsa cedars, and the monumental mood of Crown Butte.

This photographic essay, of course, is no substitute for the Lewis and Clark journals. It is, I hope, a complement to them. I saw what the men saw, and recorded with my cameras what they recorded in the journals, but sometimes it was necessary to approximate. Highways, dams, and cities are now part of the landscape.

American Odyssey is presented as a photographic essay of the broad western empire described in the journals of Lewis and Clark. The task was difficult, but pleasurable. Detailed events of the Expedition are forever lost to illustration, except by reference to the diaries, but the mentioned locations are very real, and my hope is that the curiosity aroused by these highlights will draw the reader to the original journals and their great pleasures.

Note: Certain entries from the *Original Journals of the Lewis and Clark Expedition* contain bracketed words in italic type, of which most were added by Reuben Gold Thwaites, editor of the 1904 edition. A few emendations are mine; these are bracketed and in italic and give modern names of persons and places. Those bracketed insertions in roman type were by Clark or Elliott Coues. However, to provide full enjoyment of Lewis and Clark and of the Expedition's men who kept journals, I have made no other changes in the text.

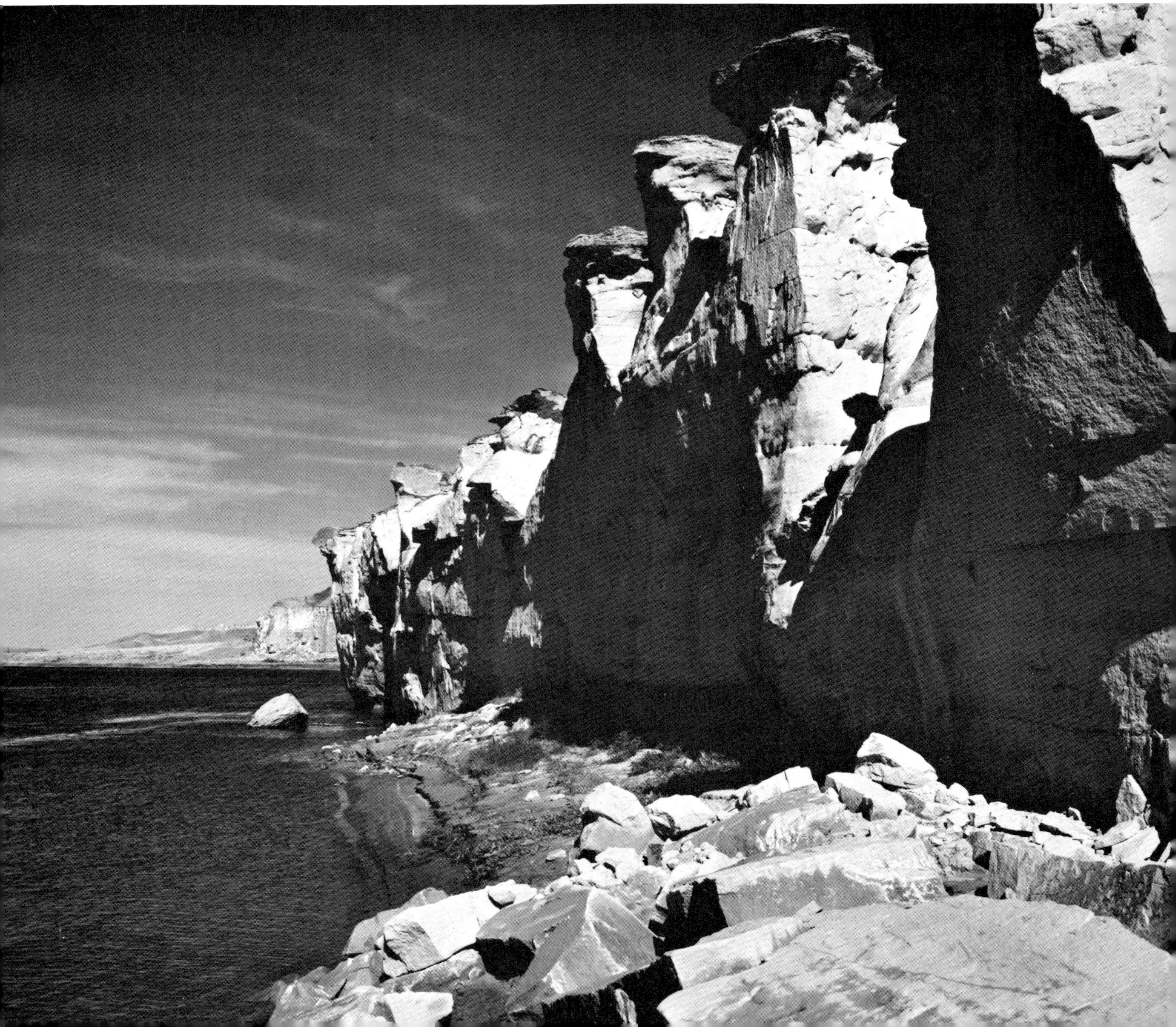

ACKNOWLEDGMENTS

My obligations to amateur and professional historians are many, and I am sincerely grateful for the assistance given me. The search for the Lewis and Clark trail across this land has been very rewarding. The many friends I have met in every section made it so. I am especially grateful to A. B. Guthrie, Jr., for his personal and unselfish enthusiasm toward my project—even long before the project started; to Ernest Staples Osgood, editor of *The Field Notes of Captain William Clark,* for his penetrating analysis of the historic mission; and to Roy E. Chatters, Pullman, Washington, whose critical wisdom influenced my perspective.

Especially helpful have been these friends at my publisher: Herbert W. Luthin, for his keen and forthright editorial appraisal; Sanford Cobb, who made me more aware of historic responsibility; and Marcia Beinhauer, for her inexhaustible patience and generosity.

The use of primary source material has been most essential to this book, and for their generous permissions I am particularly indebted to these institutions: The American Philosophical Society, Philadelphia, Pennsylvania; State Historical Society of Wisconsin (Society Press); the Filson Club, Louisville, Kentucky; University of Illinois Press, Urbana; and Yale University Press, New Haven, Connecticut.

Many persons have assisted me in research and in personal observation, and many others have accompanied me with some loss of time in their own endeavors. To these people I am indebted: Kenneth A. Keeney, E. Arnold Hanson, W. E. Steuerwald, and Del Jaquish, United States Forest Service, Northern Region, Missoula, Montana, for advice and cooperation, and to other Forest Service personnel for assistance in searching out the Lewis and Clark trail in the national forests.

Of great help, too, were R. Darwin Burroughs, Fayetteville, New York, editor of *The Natural History of the Lewis and Clark Expedition;* Paul R. Cutright, Jenkintown, Pennsylvania; E. G. (Frenchy) Chuinard, M. D., Portland, Oregon; and Eugene H. Conner, M. D., Louisville, Kentucky.

For an unforgettable 11-day trip through the Missouri River Breaks northeast of Virgelle, Montana, I thank Emil Don Tigney and his assistant, Nick Lafrantz, Havre, Montana. For additional assistance in this wild, primitive area, I am thankful for the careful planning of Orvin Fjare and former Governor Tim Babcock, Helena, Montana.

I also want to mention Robert H. Fletcher, San Diego, California; William Clark Adreon and G. Edward (Gus) Budde, St. Louis, Missouri; Ernest Ellison, D.M.D., Louisville, Kentucky; Don Munich, Jeffersonville, Indiana; Harry Hixon, Atchison, Kansas; Director Eugene Kingman and Miss Mildred Goosman, Joslyn Art Museum, Omaha, Nebraska; the late Burnby Bell, former historian at Fort Clatsop National Memorial, Astoria, Oregon; Paul Ewald, Bismarck, North Dakota; Ralph Space, Orofino, Idaho, United States Forest

Service (ret.); David Ainsworth, Salmon, Idaho, and Will G. Robinson, Pierre, South Dakota.

I am also indebted to several individuals, corporations, and organizations for their generous and courteous permission to use illustrative material: Mary Elrod Ferguson, for prints of early-day Flathead Indians; the Montana Power Company, for prints of the early-day Great Falls of the Missouri; Nebraska Game, Forestation and Parks Commission, for a photograph of a bounding deer; Oregon State Historical Society, for prints of the early-day Great Falls of the Columbia; Washington State University, Pullman, for the peace medal;

State Historical Society of North Dakota, Bismarck, for the ringed peace medal; American Philosophical Society, Philadelphia, Pennsylvania, for the drawings of Meriwether Lewis and William Clark; Arthur C. Evans, Missoula, Montana, for the Jefferson medal; United States Forest Service, for scenes of wildlife; University of Montana, Missoula, for a print of Clark's Nutcracker; Washington State Game Commission, for a beaver photograph; United States Department of the Interior, National Bison Range, Moiese, Montana, for a print of a rattlesnake and a buffalo bull; and, United States National Archives, Record Group 77, for "A Map of Lewis and Clark's Track."

Other state and national services deserve mention for their cooperation: Lewis and Clark Trail Commission, Sherry R. Fisher, chairman; Dr. Edward Crafts, director, Bureau of Outdoor Recreation; and the many local and state chairmen and individuals of the Lewis and Clark Trail Commission.

Many persons also assisted in establishing the accuracy of trails and campsites. They are: Roy E. Coy, former director of the St. Joseph, Missouri, Historical Museum, and Richard A. Wolf, director, for the use of a fine illustration of geese by D. L. Reynolds; Dr. Carl H. Chapman, University of Missouri; Kenneth R. Krabbenhoft, United States Department of the Interior, National Parks Service, Omaha, Nebraska; Alfred DuBray, Macy, Nebraska; Clark Jorgenson, Williston, North Dakota, who accompanied me to the confluence of the Yellowstone and Missouri rivers; Ray Mattison and James E. Sperry, State Historical Society of North Dakota, Bismarck; Francis O. Mitchell, Great Falls, Montana; Larry Gill, Spokane, Washington; Agnes and Jerome Vanderburg, Flathead Indians, Arlee, Montana.

It was a thrilling day when Robert H. Anderson, Cut Bank, Montana, guided me over the Two Medicine River area where Captain Lewis fought a band of the Blackfoot (Piegan) tribe. Helen B. West and Wilbur P. Werner, also of Cut Bank, provided support for my evaluation of this area, one which is increasingly gaining interest.

Others are also to be thanked: Elisabeth Walton, Parks Historian, State of Oregon; David G. Talbot, State Parks Superintendent, Salem, Oregon; Charles H. Odegaard, director, Washington State Parks & Recreation Commission; Marjorie G. Sutch, Richland, Washington, and Margaret Thompson, Kennewick, Washington; and Frans Johnson, with whom I spent several days, for locating the campsite which marked the goal of the Expedition.

It is also a genuine pleasure to give my thanks to Joe Elliott, who packed my equipment on the Missouri River Breaks, and to my typist, Edna Noelke, for her patience and dedication.

INGVARD HENRY EIDE
Missoula, Montana

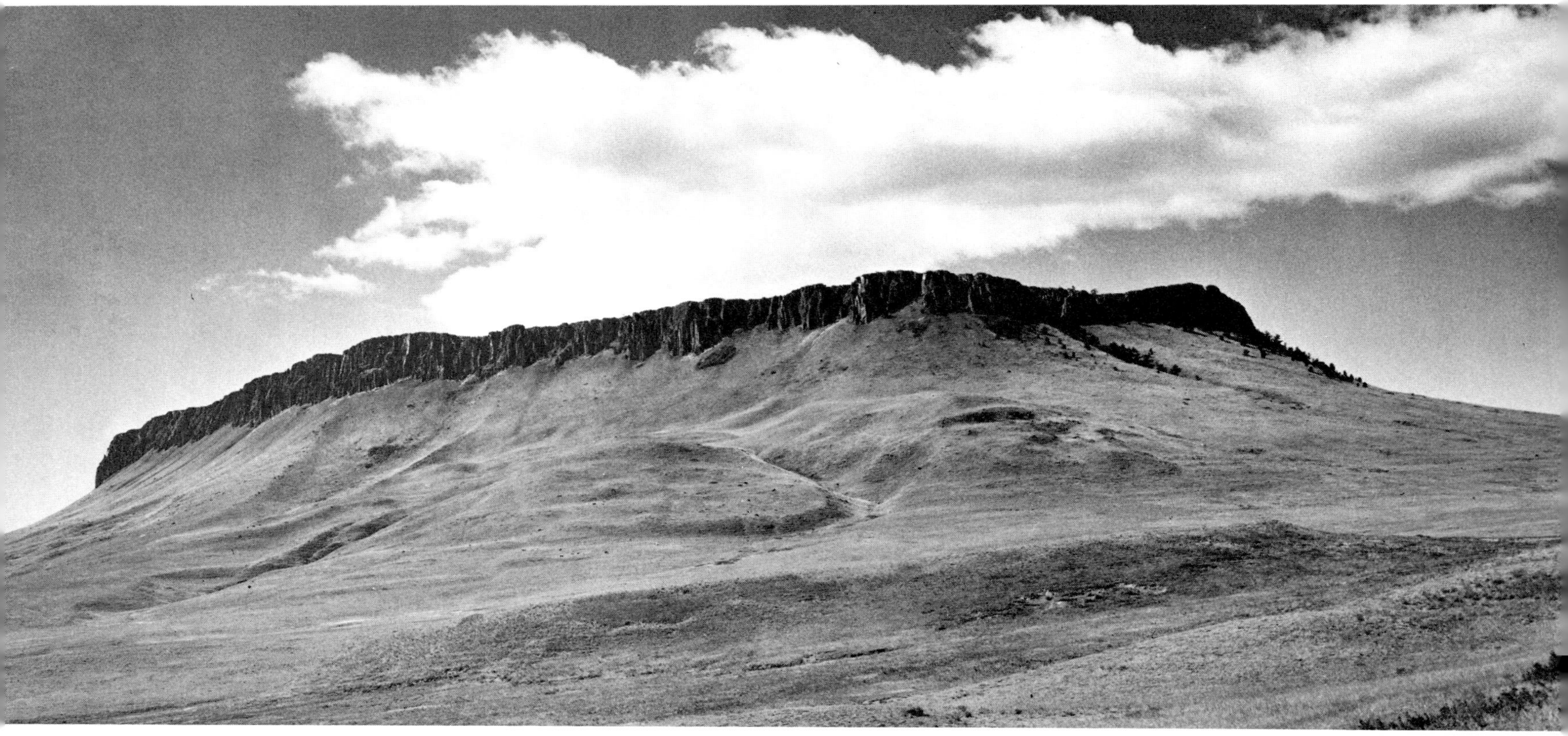

INTRODUCTION

A perceptive motion-picture executive—there are such—once told me that every good story was a love story.

He did not mean the love of man and woman necessarily, and certainly he did not imply deviation. His view comprehended mutual trust, reciprocal respect, common concern, and the ready acknowledgment between partners of both overlapping and divergent talents—an acknowledgment marked by admiration, not envy.

The definition was provocative but too broad: there are good stories outside it. Yet if ever there was a love story in his terms, it is the story of Meriwether Lewis and William Clark.

Scholars have commented on their relationship while dodging the word "love," probably because of the likelihood of misconstruction. Bernard DeVoto in the introduction to his admirable condensation of the journals says: ". . . the two agreed and worked together with a mutuality unknown elsewhere in the history of exploration and rare in any kind of human association."

"Rare" is right. Time and again and time and again the two captains, writing their separate reports, refer to each other in words not necessary to establish the success of their Expedition, words we may consider simple or naive but surely not derisive. Clark: "this point I have taken the Liberty of Calling after my particular friend Lewis." Lewis: "my friend Capt. C. is their (the Indians') favorite phisician." When Lewis fell ill, Clark saw to it that a bower of branches sheltered him from the heat. When Clark led a detachment to the Yellowstone River, heretofore seen only at its mouth, Lewis was anxious, not about his own diminished safety but the safety of his friend.

The abilities of the commanders and the circumstances attending the Expedition make the more remarkable their mutuality. They had uncommon skills in common—which doesn't insure harmony. Both were decisive, brave, and brilliant leaders and, it turned out, explorers extraordinary. One would expect friction here. There wasn't any. Nor was there friction when their individual credentials rubbed one against the other, though rubbing produces heat. Lewis was the better trained in the sciences of the time. Clark was the natural engineer and geographer and master of frontier survival. No matter. They got along. They got along even though Lewis, as the superior officer, might have exercised rank. But he had promised Clark equal authority and without strain honored the promise, though the bureaucratic War Department refused to give Clark even standing. From start to finish the two were co-captains.

Hardships such as they endured make men cranky. Day after day of exhaustion. Days of collapse and sore feet and sore-footed horses, of dawn-to-dark effort in an alien land. Days of uncertainty, as when they came to the confluence of the Marias and Missouri, and, against the hunches of the entire crew, the two leaders agreed that the right course was south. Days of no food, of survival on stinking salmon, unknown roots, horse flesh and dog meat.

Days of frustration. Days of rain and mist and of snow and of heat, of mishaps and swarming mosquitoes. Who wouldn't quarrel, one with another? Except Lewis and Clark!

The crew answered, not only to their powers of command but to their unfaltering partnership. It must be remembered that these crewmen had proficiencies of their own and with them the assurance that proficiency begets. They were individuals. They were hard cases, as they had to be to enlist and survive. An unruly outfit, one would have guessed. But only in rare instances, in the beginning weeks, the beginning miles, was assertive discipline exercised. Of necessity the measures were harsh—the lash, the gauntlet, expulsion. By them the two captains established authority. By them they made manifest their determination. Yet men such as theirs never would have continued under brute authority alone. There had to be something else, and there was—the competence acknowledged, the reciprocal trust, the mutual resolution, and the mutual dream of Lewis and Clark, which became common property. After the winter at Mandan, in present-day North Dakota, Lewis wrote of the crew, ". . . not a whisper of murmur or discontent to be heard among them . . ." The diaries of subordinates support him.

So the Expedition was united in the union of Lewis and Clark and through it united in purpose and the sense of high adventure.

What a land they saw! What an empire, stretching from the Falls of the Ohio or St. Louis—depending on your reckoning—to the misty mouth of the far Columbia! It stares at you, this country does, from old reports, made real, made here and now, by inadequate description. Inarticulation has its rightness and its eloquence. What say about a world known by no one save fractionally by aborigines? How get the feel of first-trod spaces? How express discovery? How bring home the awe of finding rivers never charted, much less named, of lifting eyes to mountains not imagined, of gazing dizzy over plains and prairies beyond the little measurements of woodsmen?

Lewis put laborious pen to paper after seeing the Great Falls of the Missouri and later on despaired, "after wrighting this imperfect description I again viewed the falls and was so much disgusted with the imperfect idea which it conveyed of the scene that I determined to draw my pen across it and begin agin, but then reflected that I could not perhaps succeed better than pening the first impression of the mind . . ."

But his lines and those of Clark, misspelled, faltering, insufficient, yet eloquent when one trades his place for theirs. It was marvels made them stumble. I have a word for the Expedition's responses to experience. The word is "wonder," a wonder beautifully reflected by Clark's repeated "butifull."

The wild Great Falls is in servitude now, a tame domestic of the Montana Power Company. Elsewhere the plow, the bulldozer, the town dump and the irrigation ditch have changed and defaced the wonderland of Lewis and Clark. But these liberties, regrettable if often warranted, are minute by comparison with those taken by the Corps of Engineers and the Bureau of Reclamation, hereafter called the Engineers, because—cite what contributions to welfare you will—both are despoilers, both the ready arms of the pork-barrelers in Washington. Neither one gives a hoot about history or knows as much about ecology as your nearest county agent.

On the great watershed of the Columbia their work has been done, maybe. On the Missouri it hasn't. Between the headwaters of Fort Peck Reservoir and Montana Power's Morony Dam there lie two hundred miles of free-flowing water, of virtually virgin stream.

What a waste! What an opportunity for the Engineers' talents. So rich was the promise that they came up with eleven proposals for development, only one of which contemplated no dams at all, though the National Park Service and other conservationists argued for a wilderness waterway. After consultations and hearings the Engineers naturally decided in favor of dams, one at Fort Benton and one at High Cow Creek. Described by that contradictory term, multipurpose, they would have almost no purpose save the production of power. Though neither has been authorized, both are being pushed, never fear, by the Engineers as well as small-town boosters who would trade history permanently in return for temporary hard cash. With the cash registers jingling, what would it matter that many a visible and cherished association with national adventure and destiny washed lost beneath the waves?

Long before the present crop of Engineers was born, Mark Twain, revisiting the Mississippi, commented on the work of the then-called United States River Commission. His words apply to the Missouri and other streams, as experience has demonstrated. "The military engineers of the Commission have taken upon their shoulders the job of making the Mississippi over again—a job transcended in size by only the original job of creating it . . . One who knows the Mississippi will promptly aver—not aloud but to himself—that ten thousand River Commissions with the mines of the world at their back, cannot tame that lawless stream, cannot curb it or confine it . . . cannot bar its path with an obstruction which it will not tear down, dance over and laugh at."

But I doubt they teach Mark Twain at the spawning grounds of Engineers.

The stretch of stream between Fort Benton and the Fort Peck Reservoir gets especial attention here not only because it is in danger of destruction but also because it is pristine, pristine beyond the lingering nature of any other reach of water ridden by the Expedition. Land passage, still subject to some argument, has been determined, charted, and marked, and its features have been celebrated by acknowledgment and enduring name, though portages and foot passes in "butifull" valleys now play host to carp and catfish. Along the dammed Missouri and along the dammed Columbia once-famous tributaries have been flooded, their mouths pushed back, or have been dammed themselves. Landmarks wait the scuba diver.

But here. But here. Here in the Missouri's breaks and badlands, among wonders ancient but made new by first enchantment.

Men besides me have made sure of camping sites and pitched camp where the Expedition camped, have seen what Lewis and Clark saw and have answered to it richly in poor words. And as it was, so is it yet. We are the captains and the crew. They turn into us. All the marvels, all the wonderments, are so much held in common that the beaver spanking water with its tail is the very one that caught attention a century and a half and more ago. A solitary cow becomes a buffalo.

If it appears from what I've said that no mark of historical importance remains save on this little stretch of the Missouri, I have gone astray. There are meadows yet and mountains, hollows and headlands, passes and pines not yet devoured, landforms and foaming creeks—all identifiable, all remindful of America's greatest journey.

Time will chip away at them, time and transmutation in the name of progress, but they remain, lying or standing or streaming like fugitives from the future; and the man who knows our history revives and reincarnates himself by seeing them.

It is because so much remains, much of it to disappear in shorter than imagined time, that the photographs by Ingvard Henry Eide became important. It hardly matters that we have not seen their like before: it matters that we may not see their like again.

Dedication makes Eide better than expert. His determination has been not only to capture in recapitulation the seasons, the very time of day, the very state of weather and the very shades of light, but also the moods that such circumstances induced in the captains and the crew. Insofar as possible, he has dodged the scarifications of civilization. By joining the Expedition, he has seen what its members saw and felt as they felt and by the magic of camera and conception has brought sight and feeling to us. How much better, how much more exciting and more moving are his pictures than mere shots!

No one may have his opportunity again, much less his ambition.

A. B. GUTHRIE, JR.

AMERICAN
ODYSSEY

THE PLAN

Washington Feb. 23. 1801.

Dear Sir [Meriwether Lewis]

The appointment to the Presidency of the U.S. has rendered it necessary for me to have a private secretary.... Your knolege of the Western country, of the army and of all it's interests & relations has rendered it desireable for public as well as private purposes that you should be engaged in that office.... Accept assurances of the esteem of Dear Sir your friend & servt.

Th: Jefferson

Pittsburgh, March 10th 1801.

Dear Sir [President Jefferson],
...that I should accept the place of your private Secretary; I most cordially acquiesce, and with pleasure accept the office....

...not a moment has been lost in making the necessary arrangements in order to get forward to the City of Washington with all possible despatch....
Your most obedient, & Very Humble Servt.,
Meriwether Lewis

April 18th 1802

Sir [Robert R. Livingston, American Minister to France]
...There is on the globe one single spot, the possessor of which is our natural and habitual enemy. It is New Orleans, through which the produce of ⅜ of our territory must pass to market, and from its [the West's] fertility it will ere long yield more than one half of our whole produce, and contain more than half of our inhabitants. France, placing herself in that door, assumes to us the attitude of defiance. The day that France takes possession of New Orleans...we must marry ourselves to the British fleet and nation....

Th: Jefferson.

[2 December 1802]

Most Excellent Señor [Pedro Cevallos, Minister of Foreign Affairs, Spain]

My Dear Sir: The President asked me the other day...if our Court would take it badly, that the Congress decree the formation of a group of travelers, who would form a small caravan and go and explore the course of the Missouri River in which they would nominally have the objective of investigating...commerce; but that in reality...the advancement of the geography. He said they would give it the denomination of mercantile, inasmuch as only in this way would the Congress have the power of voting the necessary funds.... I replied to him that...an expedition of this nature could not fail to give umbrage to our Government....

The President has been all his life a man of letters, very speculative and a lover of glory, and...he might attempt to perpetuate the fame of his administration not only by the measures of frugality and economy which characterize him, but also by discovering...the way by which the Americans may some day extend their population and their influence up to the coasts of the South Sea....

May God keep Your Excellency many years.
Washington December 2, 1802

Most Excellent Señor
Your most attentive and constant servant, kisses the hand of Your Excellency
Carlos Martínez de Yrujo
[Spanish Minister to the United States]

Confidential.   [18 January 1803]

Gentlemen of the Senate and of the House of Representatives: As the continuance of the act for establishing trading houses with the Indian tribes will be under the consideration of the Legislature at its present session, I think it my duty to communicate the views which have guided me in the execution of that act....

...An intelligent officer with ten or twelve chosen men, fit for the enterprise and willing to undertake it, taken from our posts where they may be spared without inconvenience, might explore the whole line, even to the Western Ocean, have conferences with the natives on the subject of commercial intercourse, get admission among them for our traders as others are admitted, agree on convenient deposits for an interchange of articles, and return with the information acquired in the course of two summers.
...While other civilized nations have encountered great expense to enlarge the boundaries of knowledge by undertaking voyages of discovery, and for other literary [scientific] purposes, in various parts and directions, our nation seems to owe to the same object, as well as to its own interests, to explore this, the only line of easy communications across the continent, and so directly traversing our own part of it. The interests of commerce place the principal object within the constitutional powers and care of Congress, and that it should incidentally advance the geographical knowledge of our own continent can not but be an additional gratification. The nation claiming the territory, regarding this as a literary pursuit, which it is in the habit of permitting within its dominions, would not be disposed to view it with jealousy.... The appropriation of $2,500 "for the purpose of extending the external commerce of the U.S.,"... would cover the undertaking from notice and prevent the obstructions which interested individuals might otherwise previously prepare in its way.

Th: Jefferson.

Washington Feb. 27. 1803.

Dear Sir [Dr. Benjamin Smith Barton]
...What follows in this letter is strictly confidential. You know we have been many years wishing to have the Missouri explored & whatever

river, heading with that, runs into the Western ocean. Congress, in some secret proceedings, have yielded to a proposition I made them for permitting me to have it done: it is to be undertaken immediately, with a party of about ten, & I have appointed Capt. Lewis, my secretary, to conduct it. It was impossible to find a character who to a compleat science in botany, natural history, mineralogy & astronomy, joined the firmness of constitution & character, prudence, habits adapted to the woods, & a familiarity with the Indian manners & character, requisite for this undertaking. All the latter qualifications Capt. Lewis has.... I must ask the favor of you to prepare for him a note of those in the lines of botany, zoology, or of Indian history which you think most worthy of inquiry & observation. He will be with you in Philadelphia [at the University of Pennsylvania] in two or three weeks.... Accept assurances of my friendly esteem & high respect.

Th: Jefferson

[28 February 1803]

[Passport]

The undersigned, chargé d'affaires of his Britannic Majesty in the United States of America, &c., certifies... that the bearer, Captain Merriwether Lewis... is sent (under the authority of the said United States) to explore the headwaters and shores of the Missoury and the western parts of the North American continent... to advance the scientific and literary objects of his voyage. I therefore pray all to whom these presents shall come... to render him all the aid and all the protection which shall depend upon them....

Given at the city of Washington the 28th of February 1803.

Edwd. Thornton

[1 March 1803]

[Passport]

The Commissioner General of Trade Relations chargé d'affaires of the French Republic, undersigned, requests all those to whom this present shall be delivered... to give protection and aid to the bearer, Captain Merriwether Lewis, Citizen of the United States, who... is setting out on a voyage of discovery with the purpose of exploring the Missouri river and the western regions of the Northern Continent. The undersigned certifies that Captain Merriwether Lewis has no purpose other than the above....

Given at Georgetown....

L. A. Pichon

[13 April 1803]

Dear Sir [President Jefferson]
... the future destinies of the Missouri country are of vast importance to the United States, it being perhaps the only large tract of country, and certainly the first which lying out of the boundaries of the Union will be settled by the people of the U. States.... The great object to ascertain is whether from its extent & fertility that country is susceptible of a large population....

Albert Gallatin
[Secretary of the Treasury]

Washington, April 27, 1803.

Dear Sir [Capt. Lewis]:
... I enclose you a copy of the rough draught of the instructions I have prepared for you, that you may have time to consider them & to propose any modifications which may occur to yourself as useful. ... the idea that you are going to explore the Missisipi has been generally given out; it satisfies public curiosity and masks sufficiently the real destination. I shall be glad to hear from you, as soon after your arrival in Philadelphia as you can form an idea when you will leave, & when be here. accept assurances of my constant & sincere affection.

Th: Jefferson.

[13 May 1803]

Monsieur President
I hasten to have the honor of thanking you for the letter which you have kindly sent to me by Monsieur [James] Monroe....
... I learned from your letter that you are going to have the sources of the Missouri explored, and to seek a river which, at its source, is near to the source of the Missouri, and bears its waters to the great northern ocean. This river which you wish to discover could well be the Colombia which Monsieur Gray, your fellow-citizen, discovered in 1788, or 1789. Monsieur Broughthon, one of Captain Vancouver's companions, went up that river for one hundred miles, in December 1792. He stopped at a point which he named Vancouver.... If your nation could establish an easy communication route by river, canal and short portages, between New Yorck, for example, and the town which would be built at the mouth of the Columbia, what a route that would be for trade from Europe, from Asia, and from America, whose northern products would arrive at this route by the Great Lakes and the upper Mississippi, while the southern products of the New World would arrive there by the lower Mississippi and by the Rio Norte

of New Mexico, the source of which is near the 40th parallel! What greater means to civilization than these new communication routes! . . . .

Paris, 23 floreal, year 11—13 May 1803.
B. G. E. C. Lacépède
[French naturalist]

Washington June 19th 1803

Dear Clark:

Herewith inclosed you will receive the papers belonging to your brother Genl. [George Rogers] Clark, which sometime since you requested me to procure and forward to you . . . .

From the long and uninterrupted friendship and confidence which has subsisted between us I feel no hesitation in making to you the following communication under the fulest impression that it will be held by you inviolably secret . . . .

During the last session of Congress a law was passed in conformity to a private message of the President of the United States, intiled "An Act making an appropriation for extending the external commerce of the United States." The object of this Act as understood by its framers was to give the sanction of the government to exploreing the interior of the continent of North America, or that part of it bordering on the Missourie & Columbia Rivers. . . . I am armed with the authority of the Government of the U. States for my protection, so far as its authority or influence extends; in addition to which, the further aid has been given me of liberal pasports from the Ministers both of France and England . . . . I shall embark at Pittsburgh with a party of recruits eight or nine in number, intended only to manage the boat and are not calculated on as a permanent part of my detatcment; when descending the Ohio it shall be my duty by enquiry to find out and engage some good hunters, stout, healthy, unmarried men, accustomed to the woods, and capable of bearing bodily fatigue in a pretty considerable degree: should any young men answering this description be found in your neighborhood I would thank you to give information of them on my arivall at the falls of the Ohio . . . . The present season being already so far advanced, I do not calculate on geting further than two or three hundred miles up the Missourie before the commencement of the ensuing winter. . . . You must know in the first place that very sanguine expectations are at this time formed by our Government that the whole of that immense country wartered by the Mississippi and it's tributary streams, Missourie inclusive, will be the property of the U. States in less than 12 Months from this date; but here let me again impress you with the necessity of keeping this matter a perfect secret. . . .

Thus my friend you have so far as leasure will at this time permit me to give it you, a summary view of the plan, the means and the objects of this expedition, if therefore there is anything under those circumstances, in this enterprise, which would induce you to participate with me in it's fatiegues, it's dangers and it's honors, believe me there is no man on earth with whom I should feel equal pleasure in sharing them as with yourself; I make this communication to you with the privity of the President, who expresses an anxious wish that you would consent to join me in this enterprise; he has authorized me to say that in the event of your accepting this proposition he will grant you a Captain's commission which of course will intitle you to the pay and emoluments attached to that office and will equally with myself intitle you to such portion of land as was granted to [officers] of similar rank for their Revolutionary services; the commission with which he proposes to furnish you is not to be considered temporary but permanent if you wish it; your situation if joined with me in this mission will in all respects be precisely such as my own. Pray write to me on this subject as early as possible and direct to me at Pittsburgh. . . .

With sincere and affectionate regard Your Friend & Humb^l sev^t.

Meriwether Lewis.

To Meriwether Lewis, esquire, Captain of the 1st regiment of infantry of the United States of America:

Your situation as Secretary of the President of the United States has made you acquainted with the objects of my confidential message of Jan.18, 1803, to the legislature. you have seen the act they passed, which, tho' expressed in general terms, was meant to sanction those objects, and you are appointed to carry them into execution.

Instruments for ascertaining by celestial observations the geography of the country thro' which you will pass, have already been provided. light articles for barter, & presents among the Indians, arms for your attendants, say for from 10 to 12 men, boats, tents, & other travelling apparatus, with ammunition, medicine, surgical instruments & provisions you will have prepared with such aids as the Secretary at War can yield in his department; & from him also you will recieve authority to engage among our troops, by voluntary agreement, the number of attendants above mentioned, over whom you, as their commanding officer are invested with all the powers the laws give in such a case.

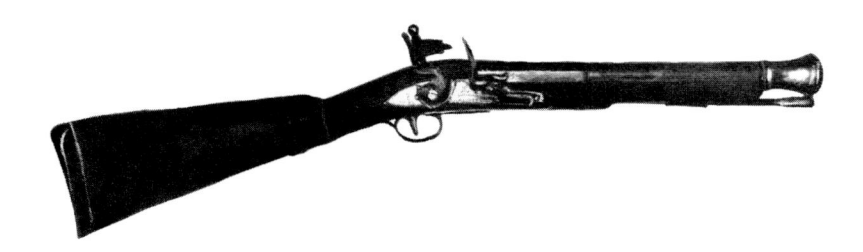

As your movements while within the limits of the U.S. will be better directed by occasional communications, adapted to circumstances as they arise, they will not be noticed here. what follows will respect your proceedings after your departure from the U.S.

Your mission has been communicated to the Ministers here from France, Spain & Great Britain, and through them to their governments: and such assurances given them as to it's objects as we trust will satisfy them. the country of Louisiana having been ceded by Spain to France, the passport you have from the Minister of France, the representative of the present sovereign of the country, will be a protection with all it's subjects: And that from the Minister of England will entitle you to the friendly aid of any traders of that allegiance with whom you may happen to meet.

The object of your mission is to explore the Missouri river, & such principal stream of it, as, by it's course & communication with the waters of the Pacific Ocean, may offer the most direct & practicable water communication across this continent, for the purposes of commerce.

Beginning at the mouth of the Missouri, you will take observations of latitude & longitude, at all remarkable points on the river, & especially at the mouths of rivers, at rapids, at islands & other places & objects distinguished by such natural marks & characters of a durable kind, as that they may with certainty be recognized hereafter. the courses of the river between these points of observation may be supplied by the compass, the log-line & by time, corrected by the observations themselves. the variations of the compass too, in different places, should be noticed.

The interesting points of the portage between the heads of the Missouri & the water offering the best communication with the Pacific Ocean should also be fixed by observation, & the course of that water to the ocean, in the same manner as that of the Missouri.

Your observations are to be taken with great pains & accuracy, to be entered distinctly, & intelligibly for others as well as yourself, to comprehend all the elements necessary, with the aid of the usual tables, to fix the latitude and longitude of the places at which they were taken, & are to be rendered to the war office, for the purpose of having the calculations made concurrently by proper persons within the U.S. several copies of these, as well as your other notes, should be made at leisure times & put into the care of the most trustworthy of your attendants, to guard by multiplying them, against the accidental losses to which they will be exposed. a further guard would be that one of these copies be written on the paper of the birch, as less liable to injury from damp than common paper.

The commerce which may be carried on with the people inhabiting the line you will pursue, renders a knolege of these people important. you will therefore endeavor to make yourself acquainted, as far as a diligent pursuit of your journey shall admit,

with the names of the nations & their numbers;
the extent & limits of their possessions;
their relations with other tribes or nations;
their language, traditions, monuments;
their ordinary occupations in agriculture, fishing, hunting, war, arts, & the implements for these;
their food, clothing, & domestic accomodations;
the diseases prevalent among them, & the remedies they use;
moral & physical circumstances which distinguish them from the tribes we know;
peculiarities in their laws, customs & dispositions;
and articles of commerce they may need or furnish, & to what extent.

And considering the interest which every nation has in extending & strengthening the authority of reason & justice among the people around them, it will be useful to acquire what knolege you can of the state of morality, religion & information among them, as it may better enable those who endeavor to civilize & instruct them, to adapt their measures to the existing notions & practises of those on whom they are to operate.

Other object worthy of notice will be

the soil & face of the country, it's growth & vegetable productions; especially those not of the U.S.

the animals of the country generally, & especially those not known in the U.S.

the remains and accounts of any which may deemed rare or extinct;

the mineral productions of every kind; but more particularly metals, limestone, pit coal & saltpetre; salines & mineral waters, noting the temperature of the last, & such circumstances as may indicate their character.

Volcanic appearances.

climate as characterized by the thermometer, by the proportion of rainy, cloudy & clear days, by lightening, hail, snow, ice, by the access & recess of frost, by the winds prevailing at different seasons, the dates at which particular plants put forth or lose their flowers, or leaf, times of appearance of particular birds, reptiles or insects.

Altho' your route will be along the channel of the Missouri, yet you will endeavor to inform yourself, by inquiry, of the character & extent of the country watered by it's branches, & especially on it's Southern side. the North river or Rio Bravo which runs into the gulph of Mexico, and the North river, or Rio colorado, which runs into the gulph of California, are understood to be the principal streams heading opposite to the waters of the Missouri, and running Southwardly. whether the dividing grounds between the Missouri & them are mountains or flatlands, what are their distance from the Missouri, the character of the intermediate country, & the people inhabiting it, are worthy of particular enquiry. The Northern waters of the Missouri are less to be enquired after, because they have been ascertained to a considerable degree, and are still in a course of ascertainment by English traders & travellers. but if you can learn anything certain of the most Northern source of the Missisipi, & of it's position relative to the lake of the woods, it will be interesting to us. some account too of the path of the Canadian traders from the Missisipi, at the mouth of the Ouisconsin river, to where it strikes the Missouri and of the soil & rivers in it's course, is desireable.

In all your intercourse with the natives treat them in the most friendly & conciliatory manner which their own conduct will admit; allay all jealousies as to the object of your journey, satisfy them of it's innocence, make them acquainted with the position, extent, character, peaceable & commercial dispositions of the U.S. of our wish to be neighborly, friendly & useful to them, & of our dispositions to a commercial intercourse with them; confer with them on the points most convenient as mutual emporiums, & the articles of most desireable interchange for them & us. if a few of their influential chiefs, within practicable distance, wish to visit us, arrange such a visit with them, and furnish them with authority to call on our officers, on their entering the U.S. to have them conveyed to this place at public expence. if any of them should wish to have some of their young people brought up with us, & taught such arts as may be useful to them, we will receive, instruct & take care of them. such a mission, whether of influential chiefs, or of young people, would give some security to your own party. carry with you some matter of the kine-pox, inform those of them with whom you may be of it' efficacy as a preservative from the small-pox; and instruct & encourage them in the use of it. this may be especially done wherever you winter.

As it is impossible for us to foresee in what manner you will be recieved by those people, whether with hospitality or hostility, so is it impossible to prescribe the exact degree of perseverance with which you are to pursue your journey. we value too much the lives of citizens to offer them to probably destruction. your numbers will be sufficient to secure you against the unauthorised opposition of individuals, or of small parties: but if a superior force, authorised or not authorised, by a nation, should be arrayed against your further passage, & inflexibly determined to arrest it, you must decline it's further pursuit, and return. in the loss of yourselves, we should lose also the information you will have acquired. by returning safely with that, you may enable us to renew the essay with better calculated means. to your own discretion therefore must be left the degree of danger you may risk, & the point at which you should decline, only saying we wish you to err on the side of your safety, & bring back your party safe, even if it be with less information.

As far up the Missouri as the white settlements extend, an intercourse will probably be found to exist between them and the Spanish posts at St Louis, opposite Cahokia, or Ste Genevieve opposite Kaskaskia. from still farther up the river, the traders *may furnish a conveyance for letters. beyond that you may perhaps be able to engage Indians to bring letters for the government to Cahokia or Kaskaskia, on promising that they shall there receive such special compensation as you shall have stipulated with them. avail yourself of these means to communicate to us, at seasonable intervals, a copy of your journal, notes & observations of every kind, putting into cypher whatever might do injury if betrayed.*

Should you reach the Pacific ocean [one full line scratched out and indecipherable] inform yourself of the circumstances which may decide whether the furs of those parts may not be collected as advantageously at the head of the Missouri (convenient as is supposed to the waters of the Colorado & Oregon or Columbia) as at Nootka sound or any other point of that coast; & that trade be consequently conducted through the Missouri & U.S. more beneficially than by the circumnavigation now practised.

On your arrival on that coast endeavor to learn if there be any port within your reach frequented by the sea-vessels of any nation, and to send two of your trusty people back by sea, in such way as shall appear practicable, with a copy of your notes. and should you be of opinion that the return of your party by the way they went will be eminently dangerous, then ship the whole, & return by sea by way of Cape Horn or the Cape of good Hope, as you shall be able. as you will be without money, clothes or provision, you must endeavor to use the credit of the U.S. to obtain them; for which purpose open letters of credit shall be furnished you authorising you to draw on the Executive of the U.S. or any of its officers in any part of the world, on which drafts can be disposed of, and to apply with our recommendations to the Consuls, agents, merchants or citizens of any nation with which we have intercourse, assuring them in our name that any aids they may furnish you, shall honorably repaid, and on demand. Our consuls Thomas Howes at Batavia in Java, William Buchanan of the isles of France and Bourbon, & John Elmslie at the Cape of good hope will be able to supply your necessities by draughts on us.

Should you find it safe to return by the way you go, after sending two of your party round by sea, or with your whole party, if no conveyance by sea can be found, do so; making such observations on your return as may serve to supply, correct or confirm those made on your outward journey.

In re-entering the U.S. and reaching a place of safety, discharge any of your attendants who may desire & deserve it, procuring for them immediate paiment of all arrears of pay & cloathing which may have incurred since their departure; & assure them that they shall be recommended to the liberality of the legislature for the grant of a soldier's portion of land each, as proposed in my message to Congress & repair yourself with your papers to the seat of government.

To provide, on the accident of your death, against anarchy, dispersion & the consequent danger to your party, and total failure of the enterprise, you are hereby authorised, by any instrument signed & written in your hand, to name the person among them who shall succeed to the command on your decease, & by like instruments to change the nomination from time to time, as further experience of the characters accompanying you shall point out superior fitness: and all the powers & authorities given to yourself are, in the event of your death, transferred to & vested in the successor so named, with further power to him, & his successors in like manner to name each his successor, who, on the death of his predecessor, shall be invested with all the powers & authorities given to yourself.

Given under my hand at the city of Washington, this 20th day of June 1803

*Th. Jefferson
Pr. U S. of America*

Washington July 2nd 1803.
Dear Mother,
The day after tomorrow I shall set out for the Western Country; I had calculated on the pleasure of visiting you before my departure but circumstances have rendered this impossible; my absence will probably be equal to fifteen or eighteen months; the nature of this expedition is by no means dangerous, my rout will be altogether through tribes of Indians who are perfectly friendly to the United States, therefore consider the chances of life just as much in my favor on this trip as I should concieve them were I to remain at home for the same length of time; the charge of this expedition is honorable to myself, as it is important to my Country. For it's fatiegues I feel myself perfectly prepared, nor do I doubt my health and strength of constitution to bear me through it; I go with the most perfect preconviction in my own mind of returning safe and hope therefore that you will not suffer yourself to indulge any anxiety for my safety....

You will find thirty dollars inclosed which I wish you to give to Sister [Jane Lewis] Anderson my love to her Edmund & the family; Reuben [Lewis] writes me that Sister Anderson has another son; remember me to Mary and Jack [Marks] and tell them I hope the progress they will make in their studies will be equal to my wishes and that of their other friends. I shall write you again on my arrival at Pittsburgh. Adieu and believe me your affectionate Son,

Meriwether Lewis

Washington, July 15, 1803.

Dear Sir [Capt. Lewis]:
   ... last night also we recieved the treaty from Paris ceding Louisiana according to the bounds to which France had a right. price 11¼ millions of Dollars, besides paying certain debts of France to our citizens which will be from 1, to 4, millions. I received also from M^r La Cepede [Comte de Lacépède], at Paris, to whom I had mentioned your intended expedition a letter .... Accept my affectionate salutations.

Th: Jefferson.

Clarksville [Ky.] 17th July 1803

Dear Lewis:
   I received by yesterday's Mail, your letter of the 19th ulto: the contents of which I received with much pleasure. The enterprise & Mission is such as I have long anticipated & am much pleased with and as my situation in life will admit of my absence the length of time necessary to accomplish such an undertaking, I will cheerfully join you in an "official character" ... and partake of the Dangers Difficulties & fatigues, and I anticipate the honors & rewards ... should we be successful in accomplishing it This is an imense undertaking fraited with numerous Dificulties, but my friend I can assure you that no man lives with whom I would prefer to undertake and share the Dificulties of such a trip than as yourself....
   I shall endeavor to engage temporally such men as I think may answer our purpose but, holding out the

Idea as stated in your letter—the subject of which has been mentioned in Louisville several weeks ago.

With every assurance of sincerity in every respect, and with aff^n y^r f^d & H. Srv.

W. C.

Clarksville 24th July 1803

Sir [President Jefferson]

I had the honor of receiving thro' Captain M. Lewis an assureance of your Approbation & wish that I would join him in a North Western enterprise. I will chearfully, and with great pleasure, join my friend Capt. Lewis in this Vast enterprise, and shall arrange my business so as to be in readiness to leave this soon after his arrival....

Wm. Clark

Louisville 24th July 1803

Dear Lewis:

... I am arrangeing my matters so as to detain but a short time after your arrival here, well convinced of the necessity of getting as far as possible up the ——————— this fall to accomplish the object as laid down by yourself and which I highly approve of.

< My friend I join you with hand & Heart and anticipate advantages which will certainly arrive from the accomplishment of so vast, Hazidous & fatiguing enterprise. You as doubt will inform the president of my determination to join you in an "official Character" as mentioned in your letter. The Credentials necessary for me to be furnished with had best be forwarded to this place, and if we set out before their arrival, to Kaskaskie. >

I have temporally engaged some men for the enterprise of a description calculated to work & go thro' those labours & fatigues which will be necessary. Several young men (gentlemen's sons) have applyed to accompany us—as they are not accustomed to labour and as that is a verry assential part of the services required of the party, I am causious in giveing them any encouragement. The newspaper accounts seem to confirm the report of war in Europe and the session of Louisiana to the United States.... Pray let me hear from you as often as possible.

Yr. W. C.

Pittsburgh August 3rd 1803.

Dear Clark: ... be assured I feel myself much gratifyed with your decision; for I could neither hope, wish, or expect from a union with any man on earth, *more perfect support or further aid in the discharge of the several duties of my mission, than that, which I am confident I shall derive from being associated with yourself....*

... if a good hunter or two could be conditionally engaged I would think them an acquisition, they must however understand that they will not be employed for the purposes of hunting exclusively but must bear a portion of the labour in common with the party....

The session of Louisiana is now no [secret]; on the 14th of July the President received the treaty from Paris, by which France has ceded to the U. States, Louisiana according to the bounds to which she had a wright, price 11¼ Millions of dollars, besides paying certain debts of France to our citizens which will be from one to four millions; the Western people may now estimate the value of their possessions.

I have been detained much longer than I expected but shall be with you by the last of this month. Your sincere friend & Ob^t. Servt.

Note—Write & direct to me at Cincinnatti

LEWIS / *August 30th 1803.*

Left Pittsburgh this day 11^{ock} with a party of 11 hands 7 of which are soldiers, a pilot and three young men on trial they having proposed to go with me throughout the voyage....

*Kaskaskias Indiana Territory
5 Sept. 1803*

Sir [brother Stephen Ordway]

... In May 1802 I received a letter from Betsey Crosby; She informed me that ... she was about offering hirself a Sacrifice at the Shrine of Hymen which information I wish to have corroborated. With respect to the flying report of a matrimonial engagement with Miss Nevens I positively deny .... not that I wish Betsey to loose an oppertunity of enjoying connubial felicity by waiting for my return; but the probability is, that if She remains in a State of celibacy till my return I may perhaps join hands with hir yet. Please to give hir my Compliments ....

*John Ordway Searj in
Russell Bissells Compy 1st Regt.*

LEWIS / *September the 7th [1803]*

Foggy this morning according to custom.... reached Wheeling 16 Miles distant .... this town is remarkable for being the point of embarkation for merchants and Emegrants who are about to descend the river [Ohio] ....

,... your ideas in the subject of judicious scelection of our party perfectly comport with my own.... there are a party of soldiers, 6 or 8 in number, now at [Fort] Massac waiting my arrival. They were scelected from the troops in the state of Tennessee... I am also authorized to scelect by voluntary engagement any men from the... posts of Massac & Kaskaskias; from these I think we shall be enabled to form our party without much difficulty; 4 or five french water-men I conceive will be essential, this we can do I presume very readily at St. Louis.

The amount of the monthly compensation(or 10$) which you have mention to the men is precisely what I have calculated on; I shall cloth and subsist the men I have with me, these will of course form a proper charge against the U. States....

Adieu and believe me your very sincere friend and associate

Meriwether Lewis

LEWIS / 15 September [1803]

... passed several bad riffles over which we were obliged to lift the boat, saw and caught by means of my dog several squirrels, attempting to swim the river... mad 18 Miles this day.

LEWIS / 18th September [1803]

The morning was clear and having had every thing in readiness the over night we set out before sunrise and at nine in the morning passed Letart's falls... this rappid is the most considerable in the whole course of the Ohio, except the rappids as they ar called opposite to Louisville in Kentuckey....

Cincinnati, September 28 1803

Dear Clark: After a most tedious and laborious passage from Pittsburgh I have at length reached this place; it was not untill the 31st of August that I was enabled to take my departure from that place owing to the unpardonable negligence and inattention of the boat builders who, unfortunately for me, were a set of most incorrigible drunkards, and with whom, neither threats, intreaties nor any other mode of treatment which I could devise had any effect; as an instance of their tardyness it may serfice to mention that they were 12 days in preparing my poles and oars.

... I am much pleased with the measures you have taken relative to the engaging the men you mention

Cincinnati, October 3rd 1803.

Dear Sir [President Jefferson]:

... So soon Sir, as you deem it expedient to promulge the late treaty, between the United States and France I would be much obliged by your directing an official copy of it to be furnished me, as I think it probable that the present inhabitants of Louisiana, from such an evidence of their having become the Citizens of the United States, would feel it their interest and would more readily yeald any information of which they may be possessed relative to the country than they would be disposed to do, while there is any doubt remaining on that subject.

... from a variety of incidental circumstances my progress has been unexpectedly delayed... I have concluded to make a tour this winter on horseback of some hundred miles through the most interesting portion of the country adjoining my winter establishment; perhaps it may be up the Canceze [Kansas] River and towards Santafee, at all events it will bee on the South side of the Missouri.... by the middle of February or 1st of March I shall be enabled to procure and forward to you such information relative to that Country, which, if it dose not produce a conviction of the utility of this project, will at least procure the further toleration of the expedition....

The water still continues lower in the Ohio than it was ever known.

I am with every sentiment of gratitude and respect, Your Obt Servt

Meriwether Lewis,
Capt. 1st U. S. Regt Infty.

The President of the United States

Kentucky Gazette and General Advertiser

[*Tuesday, November 1, 1803*]

Louisville, October 15. Captain Lewis arrived at this port on Friday last. We are informed, that he has brought barges &c. on a new construction, that can be taken in pieces, for the purpose of passing carrying-places; and that he and captain Clark will start in a few days on their expedition to the Westward.

Kentucky Gazette and General Advertiser

WASHINGTON CITY, October 22

Louisiana Treaty.

The following message was received from the President of the United States by Mr. Harvie, his secretary:

To the Senate and House of Representatives of the United States.

In my communication to you, of the 17th instant, I informed you that Conventions had been entered into, with the government of France, for the cession of Louisiana to the United States. These, with the advice and consent of the Senate, having now been ratified and my ratification exchanged for that of the First Consul of France in due form, they are communicated to you for consideration in your legislative capacity....

The ulterior provisions also suggested in the same communication, for the occupation and government of the country, will call for early attention. Such information . . . will be ready to be laid before you within a few days....

JONATHAN CLARK / 26 October 1803

Rain at Louisville at Clarksville Cap Lewis and Cap. Wm Clark set of on a Western tour....

Kentucky Gazette and General Advertiser

[*Tuesday, November 8, 1803*]

Louisville, October 29. Captain Clark and Mr. Lewis left this place on Wednesday last, on their expedition to the Westward. We have not been enabled to ascertain to what length this rout will extend, as when it was first set on foot by the President, the Louisiana country was not ceded to the United States, and it is likely it will be considerably extended—they are to receive further instructions at Kahokia. It is, however, certain that they will ascend the main branch of the Mississippi as far as possible: and it is probable they will then direct their course to the Missouri, and ascend it. They have the iron frame of a boat, intended to be covered with skins, which can, by screws, be formed into one or four, as may best suit their purposes. About 60 men will compose the party.

LEWIS / 11th November [1803]

Arived at Massac engaged George Drewyer [Drouillard] in the public service as an Indian Interpretter, contracted to pay him 25 Dollars p^r month....

JONATHAN CLARK / 12 Nov 1803

The party consisted of nine young men from Kentucky, fourteen soldiers of the United States army, who volunteered their services, two French water-men (an interpreter and hunter), and a black servant [York] belonging to Captain Clarke.

Vincennes 13th. Nov. 1803.

My dear Sir [Clark]—
...The mail of last night brought us the information that the Senate had advised the ratification of the French Treaty, 24 to 7....
I am your friend.
Willm H. Harrison

Washington, Nov. 16, 1803.

Dear Sir [Capt. Lewis]: ...I enclose you also copies of the treaties for Louisiana, the act for taking possession, a letter from Dr. [Caspar] Wister, & some information obtained by myself.... orders went from hence signed by the King of Spain & the first Consul of France, so as to arrive at Natchez yesterday evening, and we expect the delivery of the province at New Orleans will take place ... about the 25th inst.... The object of your mission is single, the direct water communication from sea to sea formed by the bed of the Missouri & perhaps the Oregon. By having mr. Clarke with you we consider the expedition double manned, & therefore the less liable to failure, for which reason neither of you should be exposed to risques by going off of your line.... As the boundaries of interior Louisiana are the high lands enclosing all the waters which run into the Missipi or Missouri directly or indirectly, *with a greater breadth on the Gulph of Mexico, it becomes interesting to fix with precision by celestial observations the longitude & latitude of the sources of these rivers, and furnishing points in the contour of our new limits.*...
... our friends & acquaintances here & in Albemarle are all well ... my friendly salutations to mr. Clarke, and accept them affectionately yourself.
Th: Jefferson.

LEWIS / Novr 16th [1803]

Passed the Missippi this day.... a respectable looking Indian offered me three beverskins for my dog [Scannon] ... of the newfoundland breed one that I prised much for his docility and qualifications generally for my journey and of course there was no bargan ... Capt. Clark and myself passed over to the ... W. side of the river from the point of junction of the rivers....

[St. Louis, 9 December 1803]

Don J. Manuel de Salcedo [Governor of Louisiana] and Marqués de Casa Calvo

The 7th of this month Mr. Merryweather Lewis, Captain of the United States army and former secretary of the President of them presented himself at this post.

...I have hinted to him that my orders did not permit me to consent to his passing to enter the Missouri River and that I was opposing it in the name of the King, my master [Charles IV].

He then told me my opinion sufficed and that from now he would not go to the said river....

I should inform Your Excellencies that according to advices, I believe that his mission has no other object than to discover the Pacific Ocean, following the Missouri, and to make intelligent observations, because he has the reputation of being a very well educated man and of many talents....

May God keep Your Excellencies many years.
Carlos Dehault Delassus

[CLARK]/ Monday the 12th of December [1803]

...nearly opposit the Messouries I came to in the Mouth of a little River Called Wood River, about 2 oClock and imediately after I had landed the NW wind which had been blowing all day increased to a Storm which was accompanied by Hail & snow ....

CLARK / Tuesday on the [Dec.] 13th [1803]

fixed on a place to build huts   Set the Men to Clearing land & Cutting Logs—a hard wind all day—flying Clouds, Sent to the Neighbourhood, Some Indians pass.

CLARK / Friday—[Dec.] 16th [1803]

Continue to raise Cabins, Sent off C Floyd to Koho [Cahokia] with Letters for Capt Lewis to put in the post office &c   Several boats pass down to day ... rais one Cabin at night   I write a Speach &c. &c.

Cahokia, December 19th 1803

Dear Sir [President Jefferson],

On my arrival at Kaskaskias, I made a selection of a sufficient number of men from the troops of that place to complete my party, and made a requisition on the Contractor to cause immediately an adequate deposit of provisions to be made at Cahokia.... This done, it became important to learn as early as possible the ultimate decision of Colo. Charles

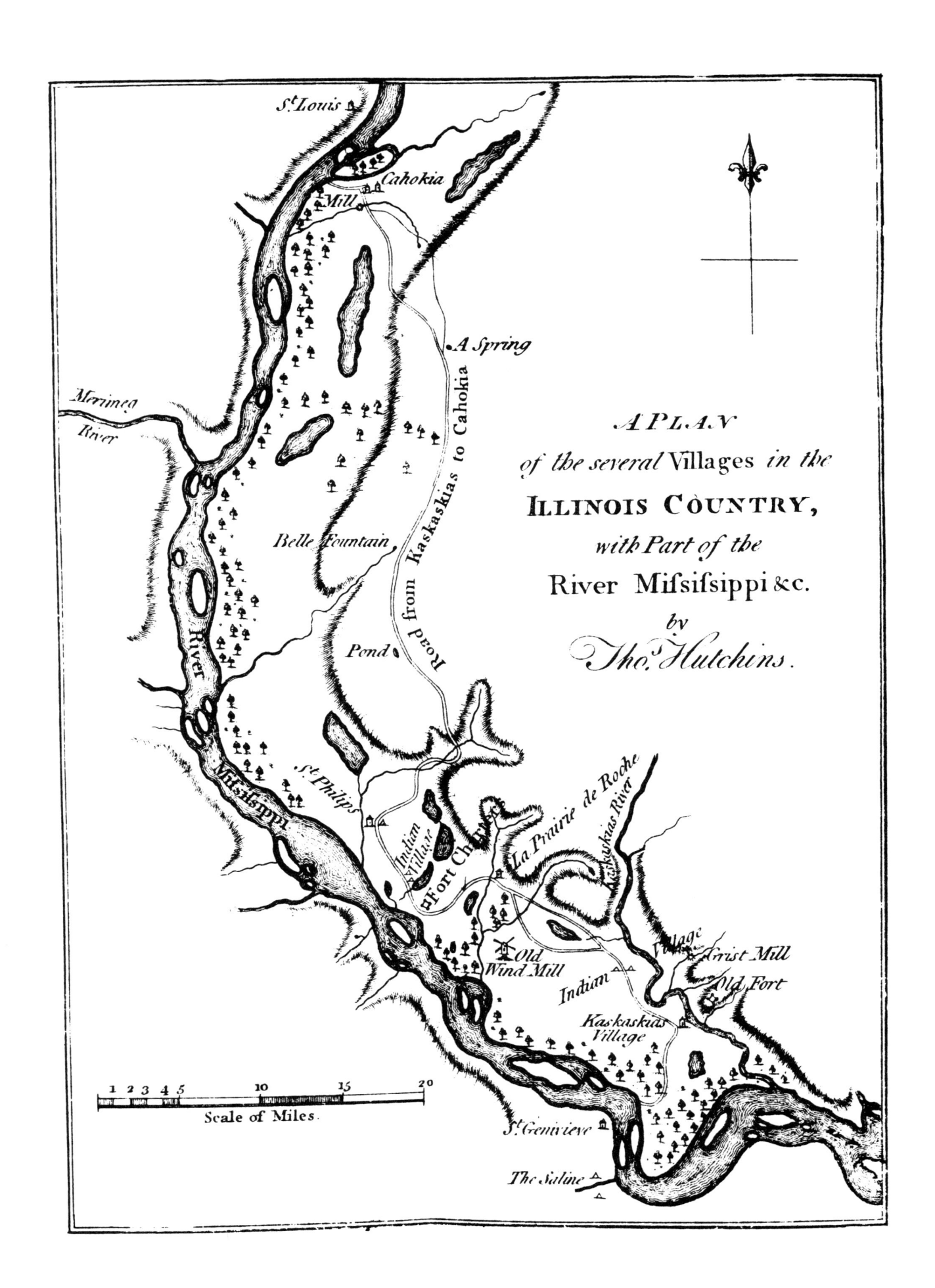

A PLAN of the several Villages in the ILLINOIS COUNTRY, with Part of the River Mifsifsippi &c. by Thos. Hutchins.

Deheau de Lassuse (the Governor of Upper Louisiana) relative to my asscending the Missouri; it became the more necessary to learn his determination on this subject, as from the advanced state of the season it must in a good measure govern my arrangements for the present winter . . . . on our arrival at his quarters we were received with much politeness by him, and . . . I proceeded to make him acquainted with the objects of my visit, handed him the Passports which I had received from the French & English Ministers . . . at the same time in a summary manner adding a few observations relative to those papers, the views of my government in fitting out this expedition, and my own wishes to proceed on my voyage . . . .

. . . I further observed, that it was not my intention at that time, to question either the policy or the right of the Spanish Government to prohibit my passage up the Missouri, that the reasons he had given for his refusal of my application, were considered by me as furnishing an ample apology on his part as an Officer for his refusal, and that I should not attempt to asscend the Missouri this season. I concurred with him in the opinion, that by the ensuing spring all obstructions would be removed to my asscending the Missouri . . . . I concluded by thanking him for the personal friendship he had evinced, in recommending to me a winter residence, which certainly in point of society or individual comfort must be considered as the most eligible of any in this quarter of the country, but that . . . I had selected for this purpose (provided it answered the description I had received of it, the mouth of a small river called Dubois on the E. side of the Mississippi opposite to the mouth of the Missouri. . . . Early the next morning Capt. Clark continuted his route with the party to the river Dubois (distant from St. Louis 18 Miles) . . . . I passed over to St. Louis with a view to obtain from the inhabitants such information as I might consider usefull to the Government, or such as might be usefull to me in my further prosecution of my voyage. I have the honor to be with much respect Your Obt. Servt.
        Meriwether Lewis Capt.
        1st U. S. Regt. Infty.

CLARK / Friday 23rd December [1803]

a raney Day   continue to put up my huts
    the men much fatigued <puting up> Carrying logs, I Send to Mr Morrisons farm for a Teem & Corn, which arived about 3 oClock, a raney Deggreeable day   Mr Griffeth Came down from his farm with a Load of Turnips &c as a present to me, Drewyear Came home to day after a <long> hunt, he Killed three Deer, & left them in the Woods, the Ice run to day   Several Deleaway [Delaware Indians] pass a chief whome I saw at Greenville Treaty, I gave him a bottle of Whiskey . . . .

CLARK / Christmas 25th Decr [1803]

I was wakened by a Christmas discharge & found that Some of the party had got Drunk <2 fought,> the men frolicked and hunted all day, Snow this morning, Ice run all day, Several Turkey Killed Shields returned with a cheese & 4 lb butter, Three Indians come to day to take Christmas with us, I gave them a bottle of Whiskey and they went off after informing me that a great talk had been held and that all the nations were going to war against the Ozous in 3 months, one informed me that a English man 16 ms from here told him that the Americans had the Countrey and no one was allowed to trade &c   I ixplaind the <thing> Intention of Govmt to him, and the Caus of the possession, Drewyear Says he will go with us . . . .

        Cahokia December 28th 1803.
Dear Sir [President Jefferson],
    On my arrival at St. Louis, the first object to which, I called my attention, was that of collecting such information as might be in some measure serviceable to . . . a wish you then entertained, if possible to induce the inhabitants of Louisiana to relinquish their landed possessions in that country, and removing with their families, accept of an equivalent portion of lands on the East side of the Mississippi, with a view more readily to induce the Indians on the East, to remove to the West side of the Mississippi, and dispose of their lands on the East side of that river to the U'States. . . .
    I have no doubt, as soon as the American government takes effect in Louisiana, that many of the best informed of it's inhabitants . . . will unsolicited come forward with much interesting information, till then, every thing must be obtained by stealth. . . .
    I am fully persuaded, that your wishes to withdraw the inhabitants of Louisiana, may in every necessary degree be affected in the course of a few years . . . .
    . . . slavery being prohibited in the Indianna Territory, (at least the further admission of any slaves), these proprietors of slaves will be compelled to deside, whether they will reside in an adjacent part of the Indianna Territory, enjoy the benefits of their indian trade, and sacrefice in some measure their slave property, or remove with these slaves to some

part of the U'States where slavery is permitted, and sacrefice all prospects of their indian trade; thus the slaves appear to me in every view of this subject to be connected with the principal difficulties with which the government will have to contend in effecting this part of it's policy....

...Your Most Obt. Servt.
Meriwether Lewis. Capt.
Ist. U.S. Regt. Infty.

CLARK / [1 Jan 1804]

Capt[s] Lewis & Clark wintered at the enterance of a Small river opposit the Mouth of Missouri Called Wood River, where they formed their party, Composed of robust helthy hardy young men ....

CLARK / January 2nd [1804]

Snow last night, <rain> a mist to day   Cap [William] Whitesides Came to See me & his Son, and some countrey people, Serjt Odderway return & bring me Some papers from Capt. Lewis, who is [in] Kohokia on business of importance to the enterprise, the party verry merry this evenig. Mr Whitesides says a no. of young men in his Neghborhood wishes to accompany Capt Lewis & myself on the Expdts Cap L. allso sent me a Letter from Capt Amos Stoddard which mentions his aptnt to the Comd of upper Louisiane, & to take possession of St. Louis &c

Washington Jan. 13. 1804.

Dear Sir [Capt. Lewis]: ... the newspapers inform us you left Kaskaskia about the 8th of December. I hope you will have recieved my letter by that day, or very soon after; written in a belief it would be better that you should not enter the Missouri till the spring; yet not absolutly controuling your own judgment formed on the spot. we have not heard of the delivery of Louisiana to us as yet, tho' we have no doubt it took place about the 20th of December, and that orders were at the same time expedited to evacuate the upper posts, troops of ours being in readiness & under orders to take possession. ... I hope this will reach you before your final departure. The acquisition of the country through which you are to pass has inspired the public generally with a great deal of interest .... present my salutations to mr. Clarke, assure all your party that we have our eyes turned on them with anxiety for their safety & the success of their enterprise. accept yourself assurances of sincere esteem & attachment.
Th. Jefferson.

January 15th 1804
River a Dubois

Dear Major [brother-in-law William Croghan]

... It is hourly expected that the American's will take possession of the other side of the Mississippi. All the Inhabitents appear anxious except the people of St. Louis, who are ingaged in the Indian Trade which they are doubtfull will be divided, amongst those whome will trade on the best terms. ...

My situation is as comfortable as could be expected in the woods, & on the frontiers; the Country back of me is butifull beyond discription; a rich bottom well timbered, from one to three miles wide, from the river to a Prarie; which runs nearly parrilal to the river from about three miles above me, to Kaskaskia and is from three to 7 miles wide, with gradual rises and several streams of runing water, and good Mill seats; This Prarie has settlements on its edges from Kahoka within three miles of this place. The Missouri which mouths imedeately opposet me <is a large turbalent> is the river we intend assending as soon as the weather will permit. This Great river which seems to dispute the preeminence with the Mississippi, coms in at right angles from the West, and forces its great sheets of muddy Ice (which is now running) against the Eastern bank....

I shall be glad to here from you at all times. Please to present my best wishes to my sisters Lucy & Fanny & the Children, to them and your self I tender the assurances [of] sincear esteem & friendship.
Wm. Clark

Washington, Jan. 22, 1804.

Dear Sir [Capt. Lewis]: ... in that of the 13th inst. I inclosed you the map of a mr Evans, a Welshman, employed by the Spanish government for that purpose, but whose original object I believe had been to go in search of the Welsh Indians said to be up the Missouri. on this subject a mr Rees of the same nation ... will write to you. N. Orleans was delivered to us on the 20th of Dec. and our garrisons & government established there. the order for the delivery of the Upper posts were to leave N. Orleans on the 28th and we presume all those posts will be occupied by our troops by the last day of the present month.... being now become sovereigns of the country, without however any diminution of the Indian rights of occupancy we are authorised to propose to them in direct terms the institution of commerce with them. It will now be proper you should inform those through whose country you will pass, or whom you may meet, that their late fathers, the Spaniards have agreed to withdraw all their troops from all the waters & country of the Missisipi & Missouri, that they have surrendered to us all their subjects Spanish and French settled there, and all their posts & lands: that henceforward we become their fathers and friends, and that we shall endeavor that they shall have no cause to lament the change: that we have sent you to enquire into the nature of the country & the nations inhabiting it, to know at what places and times we must establish stores of goods among them, to exchange for their peltries .... although you will pass through no settlements of the Sioux (except seceders) yet you will probably meet with parties of them. On that nation we wish most particularly to make a friendly impression, because of their immense power, and because we learn they are very desirous of being on the most friendly terms with us.

I inclose you a letter which I believe is from some one on the part of the Philosophical society. they have made you a member, and your diploma is lodged with me .... your friends here and in Albemarle as far as I recollect are well. ...

Accept my friendly salutations & assurances of affectionate esteem & respect.

Th. Jefferson.

CLARK / Monday 30th Jany 1804

a Cloudy morning, Some Snow ... at 8 oClock 16° abv 0, about Sun Set Capt Lewis <return> arrived accompanied by Mr J. Hay & Mr Jo Hays of Kohokia—   The hunters killed 5 Deer to day—

CLARK / Sonday 5th Feby [1804]

Still Sick, The french Man Wife &c came to See us to day   Mrs Cane als   [Samuel] Hanley sent us some Butter & milk, river riseing & Covered with Small Ice. Ct L send out Shields to get walnut Bark for pills, fowl pass

DETACHMENT ORDERS
LEWIS / Camp River Dubois, Febr 20th 1804.

The Commanding officer directs that During the absence of himself and Capt Clark from Camp, that the party shall consider themselves under the immediate command of Sergt Ordway ....

The sawyers will continue their work untill they have cut the necessary quantity of plank ....

The Blacksmiths will also continue their work ....

The four men who are engaged in making sugar will continue in that employment untill further orders ....

The practicing party will in futer discharge only one round each pr day, which will be done under the direction of Sergt Ordway, all at the same target and at the distance of fifty yards off hand. The prize of a gill of extra whiskey will be recieved by the person who makes the best shot at each time of practice.

Floyd will take charge of our quarters and store and be exempt from guard duty untill our return ....

CLARK / Feby 29 [1804]

The weather had been clear since Capt Lewis lef Camp untill this

CLARK / [March] 20th [1804]

Return from St Charles after haveing arrested the progress of a Kickapoo war party

CLARK / Monday the 26th of March 1804,

a verry smokey day   I had Corn parched to make parched meal, Workmen all at work prepareing the Boat, I visit the Indian Camps, In one Camp found 3 Squaws & 3 young ones, another 1 girl & a boy   in a 3rd Simon Gertey [Girty] & two other famileys   Gertey has the Rhumertism Verry bad ...

CLARK / Wednesday [March] 28th [1804]

a Cloudy morning, all hands at work prepareing for the voyage up the Missourie. Cap Louis arrived at 4 oClock from St Louis

CLARK / [March] 29th [1804]

Tried Several men for missconduct

CLARK / Satturday [April] 7th [1804]

Set out at 7 oClock in a Canoo with Cap Lewis my servant York & one man at ½ past 10 arrived at St. Louis. Dressed & Dined with Capt [Amos] Stoddard & about 50 Gentlemen, A Ball succeeded, which lasted untill 9 oClock on Sunday   no business to day

CLARK / Thursday [April] 19th [1804]

A rainy morning   Slept late, Thunder and lightning at 1 oClock, men Shoot at a mark ...

Camp River Dubois April the 8th 1804

Honored Parents: I now embrace this oportunity of writing to you once more to let you know where I am and where I am going. I am well thank God and in high Spirits. I am now on an expedition to the westward, with Capt Lewis and Capt Clark, who are appointed by the President of the united States to go on an Expedition through the interior parts of North America. We are to ascend the Missouri River with a boat as far as it is navigable and then go by land, to the western ocean, if nothing prevents. This party consists of 25 picked men of the armey and country likewise and I am so happy as to be one of them picked men from the armey and I and all the party are if we live to return to receive our discharge when ever we return again to the united States if we choose it. ... we expect to be gone 18 months or two years, we are to receive a great reward for this expedition 15 dollars a month and at least 400 ackers of first rate land and if we make great discoveries as we expect the united States has promised to make us great rewards, more than we are promised ....

I have received no letters since Betseys yet but will write next winter if I have a chance.

  Yours &c   John Ordway Segt.

CLARK / *Satturday [April] 28 [1804]*

Mr Hay packing up   all hands at work prepareing. Several Country men Came to Win my mens money, in doing So lost all they had, with them. river fall

St. Louis May 2nd 1804

Dear friend [Clark]: I cannot hear of or find the hair pipes. The articles you sent by Serg^t Floyd wer duly received. The mail has not arrived. The Osages will set out about the 10th ....

The pay of the men will commence from the dates of their last inlistments and will be made up to the last of November 1804 at the regular wages of soldiers & Sergt^s. &c—including the bounty of such as are intitled to it which is not the case with those whose former inlistments do not expire before the said 31st of November. Other receipt rolls will be made out for 5 dollars pr month as an advance on the score of Cloathing and provisions not furnished by the government—this to commence with those inlisted in Kentucky from the dates of their inlistments, all others from the 1st of January 1804. Mr. [Auguste] Choteau has procured seven engages to go as far as the Mandanes—but they will not agree to go further ....

  Your sincere friend
  M. Lewis   in haist

St. Louis May 6th 1804.

My dear friend [Clark],

I send you herewith inclosed your commission accompanyed by the Secretary of War's letter; it is not such as I wished, or had reason to expect; but so it is—a further explaneation when I join you. I think it will be best to let none of our party or any other persons know any thing about the grade, you will observe that the grade has no effect upon your compensation, which by G———d, shall be equal to my own.

... I send you by Colter and [Moses] Reed 200 lbs. of tallow which you will be so good as to have melted with 50 lbs. of hog's lard, cooled in small vessels and put into some of those small Keggs which wer intended for whiskey. Not a kegg can be obtained in St. Louis .... I hope all matters will be in readiness for my departure from this place. Damn Manuel [Manuel Lisa] and triply Damn Mr. B. [Francis Marie Benoit]. They give me more vexation and trouble than their lives are worth. I have dealt very plainly with these gentlemen, in short I have come to an open rupture with them; I think them both great scoundrels, and they have given me abundant proofs of their unfrendly dispositions towards our government and it's measures. These <gentlemen> (no I will scratch it out) these puppies, are not unacquainted with my opinions; and I am well informed that they have engaged some hireling writer to draught a petition and remonstrance to Govr. [William] Claibourne against me; strange indeed, that men to appearance in their senses, will <show> manifest such strong sumptoms of insanity, as to be wheting knives to cut their own throats.

I have determined to take two horses on with me, the one which is at Camp and the one the men now bring you. Adieu it is late. Your sincer friend,
  M. Lewis

CLARK / *Monday [May] 7th [1804]*

I Load the Boat all day, a fair Day   Mr Rumsey Ride a public horse to St. Louis   a fair day ...

*River a Dubois opposet the mouth of
of the Missourie River*
CLARK / *Sunday May the 13th 1804.*

I despatched an express this morning to Capt Lewis at St Louis, all our provisions Goods and equipage on Board of a Boat of 22 oars (Party) a large Perogue of 71 oares (in which 8 French) a Second Perogue of 6 oars, (Soldiers) Complete with Sails &c. &c. Men comp^d with Powder Cartragies and 100 Balls each,

all in health and readiness to set out....
Lat^d   38° - 55' - 19" - 6/10 North of equator
Long^td   89 - 57 - 45 - West of Greenwich

CLARK / *Monday May 14th 1804*

Rained the fore part of the day   I determined to go as far as St Charles a french Village 7 Leag^s up the Missourie, and wait at that place untill Capt Lewis could finish the business in which he was obliged to attend to at St Louis and join me by Land from that place 24 miles; by this movement I calculated that if any alterations in the loading of the Vestles or other Changes necessary, that they might be made at St Charles

I Set out at 4 oClock P.M, in the presence of many of the neighbouring inhabitants, and proceeded on under a jentle brease up the Missourie ....

A Journal commenced at River Dubois SGT. CHARLES FLOYD / *monday may 14th 1804*

Showery day   Capt Clark Set out at 3 oclock P m for the western expidition   the party consisted of 3 Serguntes and 38 working hands which maned the Batteow and two Perogues we Sailed up the missouria 6 miles and encamped on the N. side of the River

PVT. PATRICK GASS / [*May 14, 1804*]

On Monday the 14th of May 1804, we left our establishment at the mouth of the river du Bois or Wood river, a small river which falls into the Mississippi, on the east-side, a mile below the Missouri, and ... proceeded up the Missouri on our intended voyage of discovery, under the command of Captain Clarke. Captain Lewis was to join us in two or three days on our passage.

The corps consisted of forty-three men (including Captain Lewis and Captain Clarke, who were to command the Expedition) part of the regular troops of the United States, and part engaged for this particular enterprise. The expedition was embarked on board a batteau and two periogues. The day was showery, and in the evening we encamped on the north bank six miles up the River. Here we had leisure to reflect on our situation, and the nature of our engagements: and, as we had all entered this service as volunteers, to consider how far we stood pledged for the success of an expedition, which the government had projected; and which had been undertaken for the benefit and at the expence of the Union; of course of much interest and high expectation.

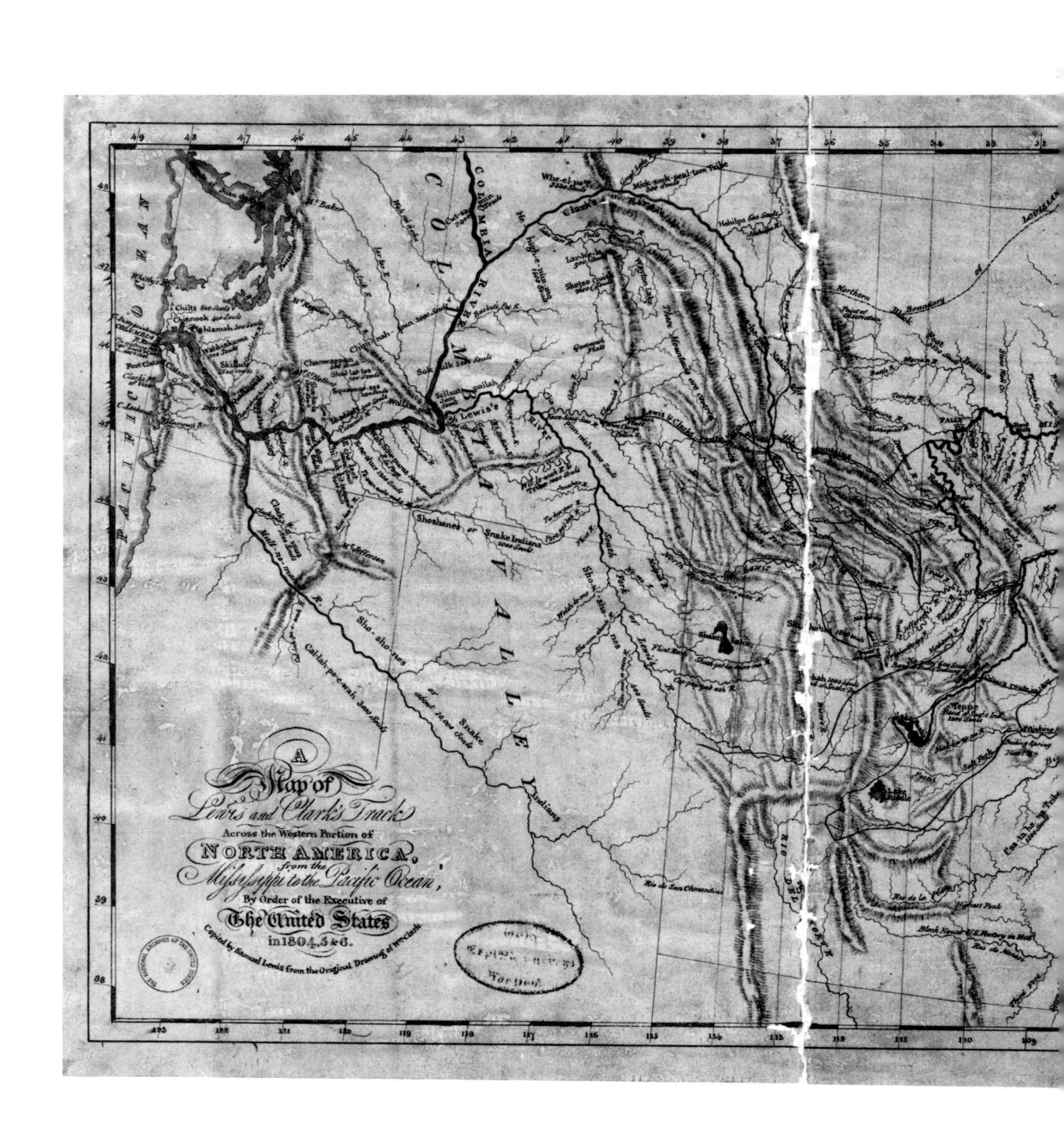

A Map of Lewis and Clark's Track Across the Western Portion of NORTH AMERICA, from the Mississippi to the Pacific Ocean; By Order of the Executive of The United States in 1804, 5 & 6. Copied by Samuel Lewis from the Original Drawing of Wm. Clark.

THE JOURNEY

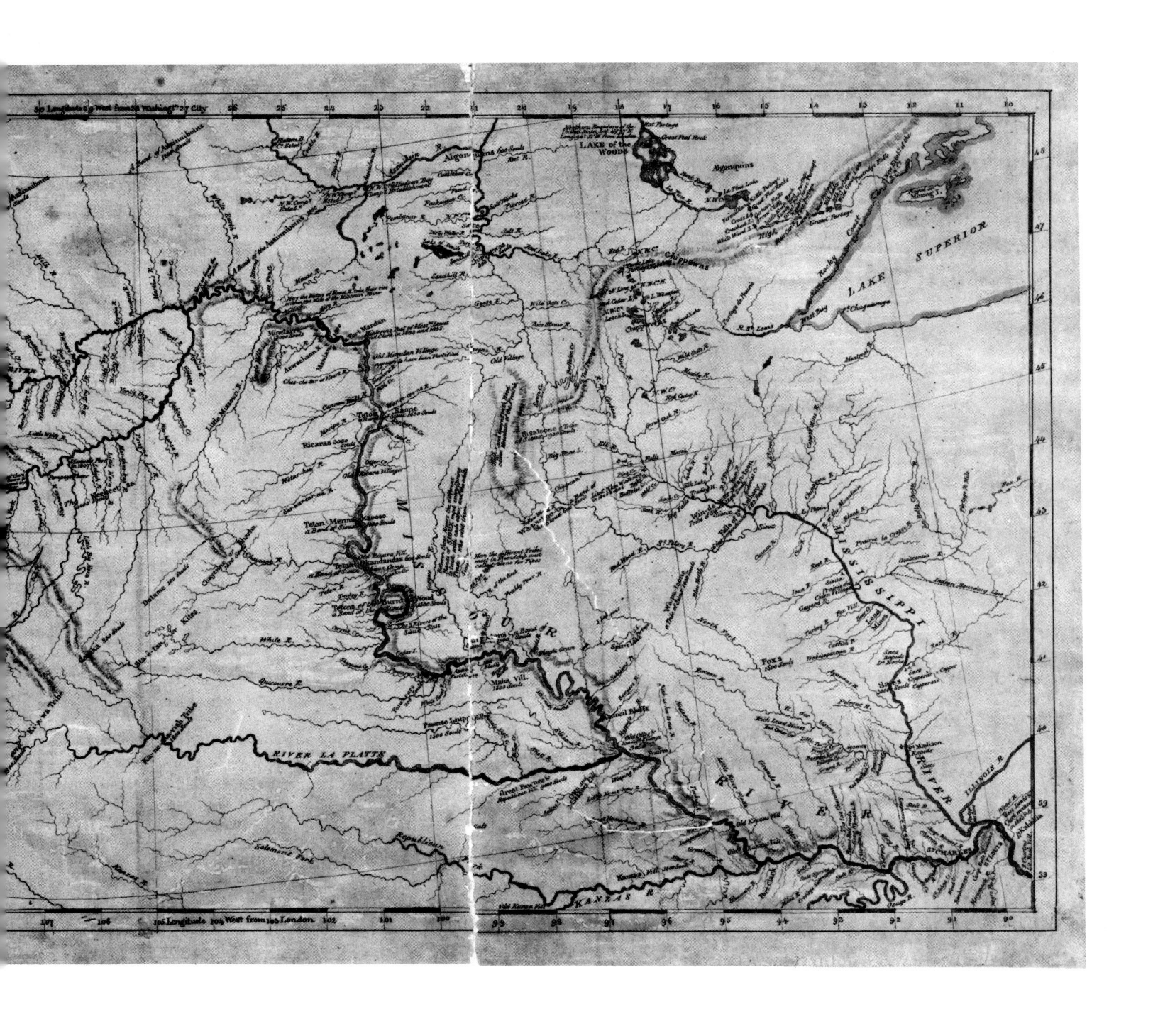

ST. CHARLES

CLARK / *May 16th Wednesday* [1804]

A fair morning Set out at 5 oClk .... we arrived at St Charles at 12 oClock a number Spectators french & Indians flocked to the bank to See the party. This Village is about one mile in length, Situated on the North Side of the Missourie at the foot of a hill from which it takes its name Peetiete Coete [petite côte] or the Little hill .... I was invited to Dine with a Mr Ducett [Duquette], this gentleman was once a merchant from Canadia, from misfortunes aded to the loss of a Cargo, Sold to the late Judge Turner he has become Somewhat reduced, he has a Charming wife an elegent Situation on the hill Serounded by orchards & a excellent gardain.

PVT. JOSEPH WHITEHOUSE / [Wedne]sday *16th May 1804.*

... this place is an old french village Situated on the North Side of the Missourie and are dressy polite people and Roman Catholicks.

WHITEHOUSE / *Friday 18th May 1804.*

... passed the evening verry agreeable dancing with the french ladies, &c.

LEWIS / *Sunday May 20th 1804*

The morning was fair, and the weather pleasent .... after bidding an affectionate adieu to my Hostis, that excellent woman the spouse of Mr Peter Chouteau, and some of my fair friends of St Louis, we set forth to that vilage in order to join my friend companion and fellow labourer Capt. William Clark, who had previously arrived at that place with the party destined for the discovery of the interior of the continent of North America ....

... we arrived at half after six and joined Capt Clark, found the party in good health and sperits. Suped this evening with Monsr Charles Tayong a Spanish Ensign & late Commandant of St Charles at an early hour I retired to rest on board the barge. St Charles is situated on the North bank of the Missouri 21 miles above it's junction with the Mississippi, and about the same distance N. W. from St Louis; it is bisected by one principal street about a mile in length runing nearly parallel with the river, the plain on which it stands is narrow ... and in the rear it is terminated by a range of small hills, hence the appellation of petit Cote, a name by which this vilage is better known to the French inhabitants of the Illinois than that of St Charles. ...

CLARK / *May 21st 1804 Monday—*

All the forepart of the Day arranging our party and procureing the different articles necessary for them at this place. Dined with Mr Ducett and Set out at half passed three oClock under three Cheers from the gentlemen on the bank and proceeded on ....

May 23, 1804

CLARK / May 23rd Wednesday 1804—

We Set out early ran on a Log and detained one hour, proceeded the Course of Last night 2 miles to the mouth of a Creek [R] on the Stbd Side called Osage Womans R.... Many people Came to See us, we passed a large Cave on the Lbd Side (Called by the french the Tavern—about 120 feet wide 40 feet Deep & 20 feet high   many different immages are Painted on the Rock at this place   the Inds & French pay omage. Many names are wrote on the rock, Stoped about one mile above for Capt Lewis who had assended the Clifts which is at the Said Cave 300 feet high, hanging over the waters.... Capt Lewis near falling from the Pinecles of rocks 300 feet, he caught at 20 foot.

WHITEHOUSE / Wednesday 23rd May 1804

... we Set out 6 oClock A. m. and proceeded on verry well. passed Some Inhabitants called boons [Daniel Boone] Settlement. passd a noted place called cave tavern in a clift of rocks....

CLARK / May 24th Thursday 1804—

... The Swiftness of the Current Wheeled the boat, Broke our Toe rope, and was nearly over Setting the boat, all hands jumped out ... by the time She wheeled a 3rd Time got a rope fast to her Stern and by the means of swimmers was Carred to Shore....

Detachment Orders.
May 26th 1804

... The sergts ... are directed each to keep a seperate journal from day to day of all passing occurrences, and such other observations on the country &c as shall appear to them worthy of notice....

The day after tomorrow lyed corn and grece will be issued to the party, the next day Poark and flour, and the day following indian meal and poark; and ... provisions will continue to be issued to the party untill further orders.... no poark is to be issued when we have fresh meat on hand....

Meriwether Lewis Capt
Wm Clark Cpt

CLARK / May 31st Thursday 1804—

... a cajaux of Bear Skins and pelteries came down from the Grand Osarge, one french man, one Indian, and a squaw,   they had letters from the man Mr Choteau Sent to that part of the Osarge nation settled on Arkansa River ... the Inds not believing that the Americans had possession of the Countrey....

June 4, 1804

FLOYD / monday June 4th [1804]

... a Butifull a peas of Land as ever I saw ... Level land on both sides.... ouer hunters Kild 8 Deer Strong water   came 10 miles....

CLARK / June 4th Monday 1804—

a fair day ... passed a Small Creek at 1 ml 15 yd Wide which we named Nightingale Creek from a Bird of that discription which Sang for us all last night, and is the first of the Kind I ever heard.... Some delightfull Land, with a jentle assent about this Creek, well timbered, Oake, Ash, Walnut &c. &c....

June 7, 1804

CLARK / June 5th Tuesday 1804—

... we had a fine wind, but could not make use of it, our Mast being broke ... our Scout discov^d the fresh sign of about 10 Inds. I expect that those Indians are on their way to war, against the Osages nation probably they are the Saukees [Sauk]. ...

CLARK / June 7$\underline{th}$ Thursday 1804—

... braekfast at the Mouth of a large Creek on the S.S. of 30 yd^s wide called big Monetou .... a Short distance above the mouth of this Creek, is Several Courious paintings and carving on the projecting rock of Limestone inlade with white red & blue flint, of a verry good quallity, the Indians have taken of this flint great quantities. ...

FLOYD / Thursday 7th June 1804

... one mile past a rock on the N. Side whare the pictures of the Devil and other things   We Kild 3 Rattel Snakes ... the Land is Good well timberd &c.

CLARK / 8$\underline{th}$ of June, Friday 1804—

... This day we met 3 men on a Cajaux from the River of the Soux above the Mahar [Omaha] Nation those men had been hunting 12 Mo: & made about 900$ in pelt^s & furs they were out of Provisions and out of Powder. rained this night.

CLARK / 12$\underline{th}$ of June. Tuesday 1804

Concluded to take old Durioun [Pierre Dorion] ... back as fur as the Soux nation with a view to get some of their Cheifs to visit the Presd^t of the United S. (This man being verry confidential friend of those people, he haveing resided with the Nation 20 odd years) ....

CLARK / 15$\underline{th}$ June, Friday 1804—

Set out early and had not proceeded far e'er we wheeled on a Sawyer which was near injuring us verry much .... we were compelled to pass under a bank which was falling in, and use the Toe rope occasionally ....

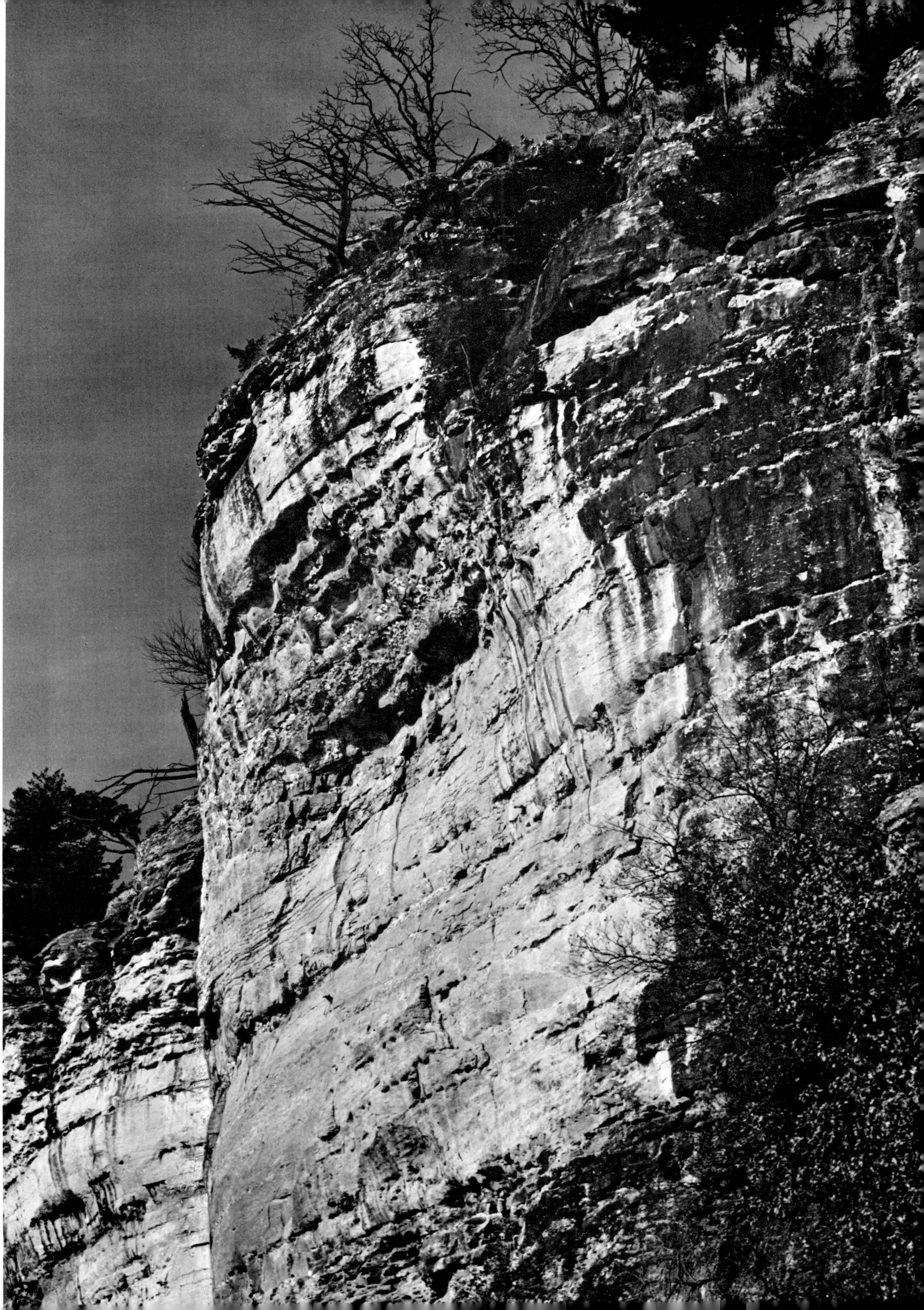

July 4, 1804

CLARK / June 17*th* Sunday 1804
(S. 65° W. 1 M*l* S. Side)—

... we Set out early and ... came too to make oars, & repair our cable & toe rope &c. ... The party is much aflicted with Boils, and Several have the Deassentary, which I contribute to the water [which is muddy.] .... The Ticks & Musquiters are verry troublesome.

CLARK / Sunday June 24*th* 1804.

... in Crossing from an Island, I got mired, and was obliged to Crawl out, a disegreeable Situation <for> & a Deverting one of any one who Could have Seen me after I got out, all Covered with mud I went to my Camp & [s]craped off the Mud and washed my Clothes ....

Camp Mouth of the Kansies
*June 29*th* 1804*

ORDERLY BOOK, CLARK

Ordered—*A Court Martiall will Set this day at 11 oClock, to consist of five members, for the trial of* John Collins *and* Hugh Hall, *Confined on Charges exhibited against them by Sergeant Floyd ....*

The Court Convened agreeable to order and proceeded to the trial of the Prisoners Viz

John Collins *Charged "with getting drunk on his post this Morning out of whiskey put under his charge as a Sentinal, and for Suffering* Hugh Hall *to draw whiskey out of the Said Barrel intended for the party."*

To this Charge the prisoner plead not Guilty.

*The Court after mature deliv[b]eration on the evidence adduced &*c* are of oppinion that the prisoner is Guilty of the Charge exibited against him, and do therefore sentence him to receive one hundred Lashes on his bear Back.*

Hugh Hall *was brought before the Court Charged with takeing whiskey out of a Keg this morning which whiskey was stored on the Bank (and under the Charge of the Guard) Contrary to all order, rule, or regulation."*

To this Charge the prisoner "Pleaded Guilty."

The Court find the prisoner Guilty and Sentence him to receive fifty Lashes on his bear Back. ...

CLARK / 30*th* June Satturday 1804

... came to at 12 oClock & rested three hours, the [day] being hot the men becom verry feeble, Farn*ts* Thermometer at 3 oClock stood 96° above 0 .... Broke our Mast

CLARK / July 4*th* Wednesday,—[1804]

ussered in the day by a discharge of one shot from our Bow piece, proceeded on .... Passed a Creek 12 yd*s* wide on L. S. comeing out of an extensive Prarie reching within 200 yards of the river, as this Creek has no name, and this being the 4*th* of July the day of the independance of the U S. call it 4*th* of July 1804 Creek .... Cap*t* Lewis walked on Shore above this Creek and discovered a high Mound from the top of which he had an extensive View .... Saw great numbers of Goslings to day which Were nearly grown, the before mentioned Lake is Clear and contain great quantities of fish and Gees & Goslings, The great quantity of those fowl in this Lake induced me to Call it the Gosling Lake, a Small Creek & several Springs run in to the Lake on the East Side from the hills   the land on that Side verry good. ... We came to and camped in the lower edge of a Plain where the 2*d* old Kanzas village formerly Stood,

July 4, 1804

above the mouth of a Creek 30 yd^s wide   this Creek we call Creek Independence as we approached this place the Prarie had a most butifull appearance

Hills & Valies intersps^d with Coops [Copses] of Timber gave a pleasing deversity to the Senery. . . . at this place the Kanzas Indians formerly lived . . . we closed the [day] by a Descharge from our bow piece, an extra Gill of whiskey.

FLOYD / *Wensday July 4th 1804*

. . . a Snake Bit Jo. Fieldes on the Side of the foot which Sweled much apply Barks to Coor [cure] . . . we camped at one of the Butifules Praries I ever Saw open and butifulley Divided with Hills and vallies all presenting themselves

July 4, 1804

July 7, 1804

July 12, 1804

SGT. JOHN ORDWAY / *Friday July 6th 1804.*

we set out eairly this morning ... the weather is verry warm, several days   the Sweet pores off the men in Streams .... a whiper will perched on the Boat for a short time.

CLARK / *July the 7th Satturday 1804—*

... passed Some Swift Water, which obliged us to draw up by roapes, a Sand bar at the point; opposit a butifull Prarie on the S. Side call^d St Michul, those Praries on the river has verry much the appearence of farms from the river   Divided by narrow Strips of woodland, which wood land is Situa^d on the runs leading to the river.... Saw a large rat on the bank. Killed a Wolf.... one man verry sick, Struck with the Sun, Capt Lewis bled him & gave Niter which has revived him much ...

CLARK / *July 12th Thursday 1804—*

Concluded to Delay here to day with a view of takeing equal altitudes & makeing observations as well as refreshing our men who are much fatigued. after early Brackfast I with five men in a Perogue assended the River Ne-Ma-haw about three Miles to the Mouth of a Small creek ... I observed artifical Mounds (or as I may more justly term graves) which to me is a strong evidence of this country being once thickly Settled....

July 12th 1804.

ORDERLY BOOK, LEWIS AND CLARK

The Commanding officers ... constituted themselves a Court Martial for the trial of such prisoners as are Guilty of Capatal Crimes, and under the rules of War punishable by DEATH.

Alexander Willard was brought foward Charged with "Lying down and Sleeping on his post" whilst a Sentinal, on the Night of the 11th Instant" (by John Ordway Sergeant of the Guard)

To this Charge the prisoner pleads Guilty of Lying Down, and Not Guilty, of Going to Sleep.

The Court after Duly Considering the evidence aduced, are of oppinion that the Prisoner Alex^{dr} Willard is guilty of every part of the Charge .... do Sentience him to receive One hundred lashes, on his bear back, at four different times ....

WHITEHOUSE / *Thurs^{dy} [July] 19th [1804]*

... as we came along Shore there was two large Cat fish had hold of Each other ... one of the french men Shot the two the first Shot.... the men pull^d a Great Quantity of wild Cherrys put them in the Barrel of whisky. Roe^d 12 Miles ....

CLARK / *July 21st Satturday 1804—*

Set out early under a gentle breeze from the S. E. proceeded on verry well ... This Great [Platte] river being much more rapid that the Missourie forces its Current against the opposite Shore. The Current of this river comes with great velocity roleing its Sands in the Missouri .... we found great dificuelty in passing around the Sand at the Mouth of this River....

July 21, 1804

July 30, 1804

*Camp White Catfish Nine [10]
Miles above the Platt River,*
CLARK / *Monday the 23ᵈ of July 1804—*

A fair morning   Set a party to look for timber for Ores, two parties to hunt, at 11 oClock Sent off George Drewyer & Peter Crousett with some tobacco to invite the Otteaus if at their town and Panies [Pawnees] if they saw them, to come and talk with us at our Camp . . . . raised a flag Staff   Sund and Dryed our provisions &c. I commence Coppying a Map of the river below to Send to the P. [President] U.S.   five Deer Killed to day   one man with a tumer on his breast,   Prepared our Camp the men put their arms in order . . . .

CLARK / *Catfish which is White Camp—
26ᵗʰ of July Thursday 1804—*

the wind Blustering and hard from the South all day . . . . I opened the Tumer of a man on the left breast, which discharged half a point. . . .

CLARK / *July 30ᵗʰ Monday 1804—*

Set out this morning early   proceeded on to a clear open Prarie on the L. S. on a rise of about 70 feet higher than the bottom which is also a Prarie (both forming Bluffs to the river) of High Grass & Plumb bush Grapes &c. and situated above high water, in a small Grove of timber at the foot of the Riseing Ground between those two preraries, and below the Bluffs of the high Prarie we Came too and formed a Camp . . . .

FLOYD / *Tuesday July 31ᵗʰ 1804*

we Lay By for to See the Indianes who we expect Hear to See the Captains. I am verry Sick and Has ben for Somtime but have Recoverd my helth again the Indianes have not Come yet   this place is Called Council Bluff . . .

WHITEHOUSE / *Tusday July 31 [1804]*

the Morning was Clear   G. Druier Catched a young beavour kept him for a pet. . . .

July 31, 1804

August 11, 1804

CLARK / August 1 1804

This being my birth day I order'd a Saddle of fat vennison, an Elk fleece & a Bevartail to be cooked and a Desert of Cheries, Plumbs, Raspberres currents and grapes of a Spur qualtity.... Musquetors verry troublesom, the Praries Contain Cheres, Apple, Grapes, Currents, Raspburry, Gooseberry Hastlenuts and a great vairety of Plants & flours not common to the US.   What a field for a Botents [botanist] and a natirless [naturalist]

WHITEHOUSE / Friday [August] 3.rd [1804]

the morning was foggy   the Indians Beheav.d well while Incamp.d Neer our party   Cap.tn Lewis Brought them to a treaty after the hour of 9 Oclock there was Six of the Zottoe [Oto] Cheifs & Six of the Missueriees; he gave 3 of the head chiefs a Meaddle Each; and the Other three Commissions in the Name of the president of the U.S.   they was well content With what they Rec.d ... they was well Content in the presence of their two fathers, which was M. Lewis & W.m Clark d.o when the Articles was Opend Out they Said as long as the french had traded with [them] they Never Gave them as much as a Knife for Nothing. Got under way in the Evening   Sail.d 5 miles.

CLARK / 11.th August Satturday 1804.—

... Cap.t Lewis myself & 10 men assended the Hill on the L. S. (under which there was some fine Springs) to the top of a high point where the Mahars King Black Bird was burried 4 years ago.... about 400 of the Mahars Died with the Small Pox ...

CLARK / 18th August, Sat'day 1804.—

... the Party with the Indians [Otoes] arriv.d ... and after a Short talk we gave them Provisions to eat & proceeded to the trial of [Moses] Reed, he confessed that he "Deserted & stold a public Rifle Shot-pouch Powder & Ball" ... only Sentenced him to run the Gantlet four times through the Party & that each man with 9 Swichies Should punish him and for him not to be considered in future as one of the Party. The three principal Chiefs petitioned for Pardin for this man   after we explained the injurey such men could doe them by false representations, & explan'g the Customs of our Countrey they were all Satisfied ....

CLARK / 19th August Sunday 1804—

... At 10 oClock we assembled the Chiefs and warriors 9 in number under an owning, and Cap.

August 20, 1804

Lewis [we] explained the Speech Sent to the Nation from the Council Bluffs by M[r] Faufon [Fairfong]. The 3 Chiefs and all the men or warriors made short Speeches approving the advice & Council their great father had Sent them, and concluded by giving themselves some Credit for their acts.

We then brought out the presents and exchanged the Big horses Meadel and gave him one equal to the one Sent to the Little Thief & gave all Some Small articles & 8 Carrots of Tobacco, we gave one Small Meadel to one of the Chiefs and a Sertificate to the others of their good intentions....

CLARK / 20th August Monday 1804.—

Sergeant Floyd much weaker and no better.... as bad as he can be no pulse & nothing will Stay a moment on his Stomach or bowels.... at the first Bluff on the S. S. Serj. Floyd Died with a great deal of Composure, before his death he Said to me, "I am going away" I want you to write me a letter." We buried him on the top of the bluff ½ Mile below a Small river to which we Gave his name.... This Man at all times gave us proofs of his firmness and Determined resolution to doe Service to his Countrey and honor to himself   after paying all the honor to our Decesed brother we camped in the Mouth of floyds River about 30 yards wide, a butiful evening.

WHITEHOUSE / Thursday [August] 30th [1804]

... about 9 oClock the Indians was brought across the river in our pearogue   our Captains counseled with them read a Speech to them, & made 5 of them chiefs & Gave them all Some Marchandize &c &c. They received them verry thankfully divided them out among themselves, & play on their juze harps, Sung &c.... Cap[t] Lewis Shot his air gun told them that their was medicien in hir & that She would doe Great execution, they were all amazed at the curiosity....

CLARK / Sept. 7th Friday [1804]—

... discovered a Village of Small animals that burrow in the grown (those animals are Called by the French Petite Chien) Killed one and Caught one a live by poreing a great quantity of Water in his hole ... those little animals Set erect make a Whistleing noise and whin allarmed Step into their hole.... Those Animals are about the Size of a Small Squrel Shorter (or longer) & thicker, the head much resembling a Squirel in every respect ... they have fine fur & the longer hair is gray....

September 7, 1804

September 14, 1804

September 20, 1804

September 23, 1804

ORDWAY / *Friday 14th Sept 1804.*

... Capt Clark joined us   had killed a curious annamil resembling a Goat ... it was 3 feet high resembles a Deer in some parts   the legs like a Deer. feet like a Goat. horns like a Goat only forked .... Goat [antelope] Stuffed in order to Send back to the city of Washington. the bones and all.

CLARK / *20th of September, Thursday 1804—*

... I walked on Shore with a view of examening this bend   crossed at the Narost part which is a high irregular hills of about 180 or 190 feet, this place the gouge of the bend is 1 mile & a quarter (from river to river or across,) from this high land which is only in the Gouge, the bend is a Butifull Plain thro which I walked ... Camped late on a Sand Bar near the S.S. [starboard side] ...

WHITEHOUSE / *Thursday 20th Sept [1804]*

... at 2 oC we proceeded on passed a long range of bluffs on N. S. of a dark coulour. out of those and others of the same kind is where the Missourie Gets its muddy colour for this Earth melts like Sugar ....

CLARK / *23rd of September Sunday 1804—*

... the river is nearly Streight for a great distance wide and Shoal (4) passed a Creek on the S.S. 16 yards wide we Call Reuben Creek, as R. Fields found it. Camped on the S. S. below the mouth of a Creek on the L. S. ...

CLARK / *24th September Monday 1804—*

... The Tribes of the Seauex Called the Teton, is Camped about 2 Miles up on the N.W. Side, and we Shall Call the River after that Nation, Teton This river is 70 yards wide at the mouth of Water, and has a considerable Current   we anchored off the Mouth. ...

    ... we prepare to Speek with the Indians tomorrow ....

October 18, 1804

October 19, 1804

CLARK / 25th Sept. [1804]—
... Envited those Cheifs on board to Show them our boat and such Curiossities as was Strange to them, we gave them ¼ a glass of whiskey which they appeared to be verry fond of, Sucked the bottle after it was out & Soon began to be troublesom, one of the 2d Cheif assumeing Drunkness, as a Cloake for his rascally intentions   I went with those Cheifs ... to Shore with a view of reconsileing those men to us, as Soon as I landed the Perogue three of their young Men Seased the Cable of the Perogue, (in which we had pressents &c) the Chiefs Sold.r [each Chief has a soldier] Huged the mast, and the 2d Chief was verry insolent both in words & justures ... I felt My self Compeled to Draw my Sword (and Made a Signal to the boat to prepare for action) at this Motion Capt Lewis ordered all under arms in the boat, those with me also Showed a Disposition to Defend themselves and me, the grand Chief then took hold of the roap & ordered the young Warrers away, I felt My Self warm & Spoke in verry positive terms.

Most of the Warriers appeared to have ther Bows strung and took out their arrows from the quiver. as I (being surrounded) was not permited (by them) to return, I sent all the men except 2 Inps [Interpreters] to the boat, the perogue Soon returned with about 12 of our determined men ready for any event. this movement caused a no: of the Indians to withdraw at a distance, (leaving their chiefs & soldiers alone with me). ... I turned off & went with my men on board the perogue, I had not prosd more the 10 paces before the 1st Cheif 3rd & 2 Brave Men Waded in after me. I took them in & went on board ....

WHITEHOUSE / Tuesday 25th Sept [1804]
... Capt Clark told them that he had men and medican on board that would kill 20 Such nations in one day. they then began to be Still and only wished that we would Stop at their lodges untill their women & children would see us. ...

WHITEHOUSE / Wednesday 26th Sept 1804
... the natives appeared peacable & kind. ... when the Indians Saw the officers comming they Spread a buffaloe Robe on the Ground and they Set down on it, then it was taken up by 4 warriers and carried to the Grand chiefs lodge. they killed Several fat dogs which they call the best meat that ever was.

CLARK / 10th of October Wednesday 1804.
... Those Indians [Arikara] wer much astonished at my Servent, they never Saw a black man before, all flocked around him & examind him from top to toe, he Carried on the joke and made himself more turribal than we wished him to doe. ...

CLARK / 13th of October Satturday 1804—
one man J. Newmon [John Newman] confined for mutinous expression .... a fiew miles from the river on the S.S. [starboard side] 2 Stones resembling humane persons & one resembling a Dog is Situated in the open Prarie, to those Stones the Rickores [Arikara] pay Great reverance make offerings (votive Dress &c.) whenever they pass (Informtn of the Chief & Intepeter) those People have a curious Tredition of Those Stones, one was a man in Love, one a Girl whose parents would not let marry [The Man as is customary went off to mourn, the female followed.), the Dog went to morn with them   all turned to Stone gradually, commenceing at the feet. Those people fed on grapes untill they turned, & the woman has a bunch of grapes yet in her hand ....

we Tried the Prisoner Newmon ... by 9 of his Peers they did "Centence him 75 Lashes & Disbanded [from] the party."

CLARK / 18th of October Thursday 1804—
Set out early proceeded on at 6 mls passed the mouth of la [Le] Boulet (or Cannon Ball River) about 140 yards wide on the L.S.   this river heads in the Court Noi or Black Mountains (a fine Day) above the mouth of the river Great numbers of Stone perfectly round with fine Grit are in the Bluff and on the Shore, the river takes its name from those Stones which resemble Cannon Balls. ...
... The Countrey in this Quarter is Generally leavel & fine   Some high Short hills, and some ragid ranges of Hills at a Distance

CLARK / 19th October Friday 1804.—
... I walked out on the Hills & observed Great numbers of Buffalow .... I saw Some remarkable round hills forming a cone at top one about 90 foot one 60 & several others Smaller, the Indian Chief say that the Callemet bird [golden eagle] live in the holes of those hills ....

October 19, 1804

CLARK / 21st October Sunday 1804

a verry Cold night wind hard from the N.E. . . . Some distance up this River [Chiss-che-tar, the Heart River] is Situated a Stone which the Indians have great faith in & say they See painted on the Stone, all the Calemetes & good fortune to hapin the nation & parties who visit it. a tree [an oak] which Stands [alone] near this place [about 2 miles off] in the open prarie . . . they [Mandans] pay Great respect to   make Holes and tie Strings thro [the skins of their] their necks and around this tree to make them brave . . . at 2 Miles passed the 2nd Village of the Mandins which was in existance at the same time with the 1st   this Village is at the foot of a hill on the S. S. in a butifull & extensive plain . . . at this time covered with Buffalow. . . .

CLARK / 22nd October Monday 1804—

last night at 1 oClock I was violently and Suddenly attacked with the Rhumetism in the neck which was So violent I could not move   Capt [Lewis] applied a hot Stone raped in flannel, which gave me some temporey ease. . . .

CLARK / 24th October Wednesday 1804—

. . . no game on the river to day—a prof of the Indians hunting in the neighbourhood

CLARK / 25th of October Thursday 1804.—

. . . a frenchman has latterly been killed by the Indians on the Track to the tradeing establishment on the Ossinebine R. in the North of this place (or British fort) . . . Several Indians came to see us this evening . . . customary for this [Mandan] nation to Show their greaf by some testimony of pain, and that it was not uncommon for them to take off 2 Smaller fingers of the hand (at the 2d joints) and some times more with other marks of Savage effection . . .

CLARK / 26th of October Friday 1804—

. . . at this Camp Saw a (Mr) McCracken Englishmon from the N. W. (Hudson Bay) Company this man Came nine Days ago to trade for horses & Buffalow robes,—one other man came with him. the Indians continued on the banks all day. but little wood on this part of the river. . . .
. . . camped on the L.S. about ½ a mile below the 1st Mandin Town on the L.S. . . . many men womin & children flocked down to See us, Capt Lewis walked to the village with the principal Chiefs and our interpters, my Rhumatic complaint increasing I could not go.

October 21, 1804

October 26, 1804

October 27, 1804

CLARK / *27th of October Satturday 1804, Mandans.—*

We Set out arly   came too at the Village on the L. S.   this village situated on an eminance of about 50 feet above the Water in a handsom plain it containes houses in a kind of Picket work, the houses are round and verry large containing several families . . . .

[Sgt.] GASS / *Saturday [October] 27th. [1804]*
. . . This village contains 40 or 50 lodges, built in the manner of those of the Rickarees. . . . some of the children have fair hair. . . .

ORDWAY / *Saturday 27th Oct [1804]*
. . . this is the most conveneint place to hold a counsel with the whole nation. we hoisted a flag pole &. C.   from the mouth of the Missouri to this place is 1610 miles.

CLARK / *Sunday. 28th of October 1804—*
. . . entertained Several of the Curious Chiefs whome, wished to see the Boat which was verry curious to them viewing it as great medison, (whatever is mysterious or unintelligible is called great medicine) as they also Viewed my black Servent . . .

Upper Mandane Village, Oct. 31, 1804.
To Charles Chaboiller, Esq. of the N. W. Co.
Sir: . . . We have been commissioned and sent by the government of the United States for the purpose of exploring the river Missouri, and the western parts of the continent of North America, with a view to the promotion of general science. Your government have been advised of the voyage . . . .
. . . We shall, at all times, extend our protection as well to British subjects as American citizens . . . it is consonant with the liberal policy of our government, not only to admit within her territory the free egress and regress of all citizens and subjects of foreign powers with which she is in amity, but also to extend to them her protection, while within the limits of her jurisdiction. . . .
We are, with much respect, Your ob't. serv'ts.
Meriwether Lewis,
Capt. 1st U. S. R[egt.] Inf.
William Clark,
Capt. [2d.Lt. U. S. Artillerists]

CLARK / *1st of November. Thursday 1804—*
. . . Mr Mc Crackin . . . Set out at 7 oClock to the Fort on the Ossiniboin   by him Send a letter, (inclosing a Copy of the British Ministers protection [passport]) to the principal agent [Charles Chaboillez] of the Company. . . .

CLARK / *3rd of November Satterday 1804—*
. . . engaged one man [Baptiste Le Page] (a Canadian Frenchman who had been with the Chayenne Ind. on the Côte noir & last summer descended thence the Little Missouri—he was of our permanent.) Set the french who intend to return [to St. Louis] to build a perogue, many Indians pass to hunt, Mr Jessomme (Jesseaume) with his Squar & children come down to live, as Interpter . . . in the evening the Ka goh ha me or little ravin came & brought us on his Squar (who carried it on her back) about 60 Wt of Dried Buffalow meat a roabe, & Pot of Meal &c.   they Delayed all night . . .

CLARK / *5 November Monday 1804—*
I rose verry early and commenced raising the 2 range of Huts [Fort Mandan]   the timber large and heavy all to carry on on Hand Sticks, cotton wood & Elm Som ash Small, our situation Sandy, great numbers of Indians pass to and from hunting . . .

ORDWAY / Monday 5th No^v 1804.

... all hands ... Splitting out punchiens for to lay the loft which we intend covering over with earth to make the huts more warm and comfortable. ...

CLARK / 9th Nov. Friday 1804—

... Horses Dogs & people all pass the night in the Same Lodge or round House, Cov^d with earth with a fire in the middle ....

CLARK / 11th November Sunday 1804. Fort Mandan

a cold Day   continued at work at the Fort ... two squars [one, Sacajawea] of the Rock mountains, purchased from the Indians by a frenchmen (Chaboneau) came down ...

CLARK / 18th Nov. Sunday 1804—

a cold morning Some wind   the Black Cat, Chief of the Mandans came to see us, he made great inquiries respecting our fashions, he also Stated the Situation of their nation ... we advised them to remain at peace & that they might depend upon Getting Supplies through the Channel of the Missourie, but it required time to put the trade in opperation. ...

CLARK / 22nd of November Thursday 1804—

... I was allarmed about 10 oClock by the Sentinal, who informed that an Indian was about to kill his wife .... the Husband observed that one of our Serjeants Slept with his wife ... We derected the Serjeant (Odway) to give the man Some articles ... and advised him to take his squar home and live hapily together in future ...

ORDWAY / Thursday 22^d No^v [1804]

pleasant and warm .... the pearogue returned towards evening with ab^t 12 bushels of mixed coullour^d corn in ears which the natives took out of the Ground where they burry it in holes in their village.

CLARK / 16th December Sunday 1804—

... the Thermt^r at Sun rise Stood at 22° below 0 .... Mr. Henny from the [Hudson's Bay] Establishment on River Ossinniboin ... arrived ....

GASS / Sunday [December] 16th. [1804]

... The object of the visits we received from the N. W. Company, was to ascertain our motives for visiting that country, and to gain information with respect to the change of government.

CLARK / 17th December Monday 1804—

... We found Mr. Henny a Verry intelligent Man from whome we obtained Some Scetches of the Countrey between the Mississippi & Missouri, and Some Sketches ... obtained from the Indin^s to the West of this place also the names and charecktors of the Seeaux &c.   about 8 oClock PM. the thermometer fell to 74° below the freesing pointe.

WHITEHOUSE / Monday 17 Dec^r [1804]

a cold day. Serg^t Gass fixed a horse Sled for one of of the N. W. Comp^y traders to go to thier forts ....

CLARK / 21st December Friday 1804—

... the Indian whome I stoped from Commiting Murder on his wife, 'thro jellosy of one of our interpeters, Came & brought his two wives and Shewed great anxiety to make up with the man with whome his joulussey Sprung. ...

ORDWAY / Tuesday 1st Jany 1805.

... 15 of the party went up to the 1st village of Mandans .... carried with us a fiddle & a Tambereen & a Sounden horn. ... a frenchman danced on his head and all danced round him for a Short time ...

November 18, 1804

January 7, 1805

CLARK / 5th of January Satturday 1805—

a cold day Some Snow ... I imploy my Self Drawing a Connection of the Countrey from what information I have recved. a Buffalow Dance ... for 3 nights passed in the 1st Village, a curious Custom the old men arrange themselves in a circle & after Smoke a pipe ... the young men who have their wives back of the Circle go [each] to one of the old men with a whining tone and request the old man to take his wife (who presents [herself] necked except a robe) and—(or Sleep with her) the Girl then takes the Old Man (who verry often can scarcely walk) and leades him to a convenient place for the business, after which they return to the lodge ... (We Sent a man to this Medisan Dance last night, they gave him 4 Girls) all this to cause the buffalow to Come near So that they may Kill them

CLARK / 7th of January Monday 1805—

... the river fell 1 inch ... the 3 men returned from hunting, they kill^d, 4 Deer .... I continue to Draw a connected plott from the information of Traders, Indians & my own observation & ideas. from the best information, the Great falls is about (800) miles nearly West,

January 13, 1805

February 5, 1805

Clark / *13th of January Sunday 1805*

a Cold Clear Day (great number of Indians move Down the River to hunt)   those people Kill a Number of Buffalow near their Villages and Save a great perpotion of the Meat, theer Custom of makeing this article of life General leaves them more than half of their time without meat   Their Corn & Beans &c they keep for the Summer, and as a reserve in Case of an attack from the Soues, [of] which they are always in dread .... Chaboneu informs that the Clerk of the Hudson Bay Co. with the Me ne tar res has been Speaking Some fiew express*ns* unfavourable towards us ....

Ordway / *Monday 14th Jany [1805]*

... G. Shannon came in this evening and informed us that Whitehouse had his feet frost bit & could not come in without a horse ...

Whitehouse / *Wednesday 16th Jany 1805.*

quite warm for the time a year & pleasant   the Snow melted fast.   I came to the fort & 2 more men with me   my feet got Some easier.

Clark / *16th January Wednesday 1805*

... one of the 1st War Chiefs of the big bellies [Gros Ventres] nation Came to see us to day with one man and his Squar to wate on him (requested that she might be used for the night) (his wife handsome) We Shot the Air gun, and gave two Shots with the Cannon which pleased them verry much ...

This War Chief gave us a Chart in his Way of the Missourie, he informed us of his intentions of going to War in the Spring against the Snake Indians   we advised him to look back at the number of Nations who had been distroyed by War, and reflect upon ...

if he went to War against those Defenceless people, he would displease his great father, and he would not receive that pertection & care from him as other nations who listened to his word. This Chief who is a young man 26 y.r old replied that if his going to war ... would be displeasing to us he would not go, he had horses enough. ...

CLARK / 19*th* January Satturday 1805.

a fine Day   Mess.rs Le rock & M.cKinzey returned home ...

WHITEHOUSE / Saturday 19*th* Jan.y 1805.

2 men Sent with three horses down the River for meat to the hunting Camps, which is about 30 miles distant from the Fort. the way they go [is] on the Ice.

WHITEHOUSE / Sunday 20*th* Jan.y 1805.

Some men went up to the villages. ... when they had done eating they [Mandans] gave a bowl of victuls to a buffalows head which they worshiped, & S.d Eat this So that the live ones may come in that we may git a Supply of meat. Some of them & indeed the most of them have Strange & uncommon Ideas, but verry Ignorant of our forms & customs, but quick & Sensible in their own way & in their own conceit &c &c.

CLARK / 27*th* of January Sunday 1805

a fine day, attempt to Cut our Boat and Canoos out of the Ice, a deficuelt Task I fear as we find water between the Ice, I bleed the man with the Plurisy to day & Swet him, Cap.t Lewis took off the Toes of one foot of the Boy who got frost bit Some time ago ...

GASS / Tuesday [January] 29th. [1805]

We attempted another plan for getting our water-craft disengaged from the ice: which was to heat water in the boats, with hot stone, but in this project we failed, as the stones we found would not stand the fire, but broke to pieces.

ORDWAY / Wednesday 30*th* Jan.y 1805.

Some cloudy. Serg.t Gass Sent up the River to another bluff in order to look for another kind of Stone that would not Split with heat   he brought one home & het it ... as soon as it was hot it bursted asunder   So we Gave up that plan.

LEWIS / 5*th* February Tuesday 1805.—

... visited by many of the natives who brought a considerable quanty of corn in payment for the work which the blacksmith [Pvt. John Shields] had done for them   they are pecuarly attached to a battle ax formed in a very inconvenient manner in my opinion. it is fabricated of iron only ....

LEWIS / 11*th* February Monday 1805.

... this evening one of the wives [Sacajawea] of Charbono was delivered of a fine boy. ... M.r Jessome informed me that he had freequently administered a small portion of the rattle of the rattle-snake ... to produce the desired effect, that of hastening the birth of the child ... he administered two rings of it to the woman broken in small pieces with the fingers and added to a small quantity of water. ... I was informed that she had not taken it more than ten minutes before she brought forth   perhaps this remedy may be worthy of future experiments, but I must confess that I want faith as to it's efficacy.

CLARK / 21*st* February Thursday 1805

... Cap.t Lewis ... after finding that he could not overtake the Soues War party, (who had in their way distroyed all the meat at one Deposit which I had made & Burnt the Lodges) ... hunted two day Killed 36 Deer & 14 Elk ... the meet which he killed and that in the lower Deposit amounting to about 3000.lb was brought up on two Slays   one Drawn by 16 men had about 2400.lb on it

ORDWAY / Thursday 28*th* Feb.y 1805.

... the Souix Savvages who Robed our men of the 2 horses ... was 106 in nomber and ... they had a mind for to kill our men & that they held a counsel .... but while they were doing that our men were off and got clear .... M.r Tabbo [Antoine Tabeau, Arikara trader] a frenchman ... Sent a letter up to the commanding officers & Mandan chiefs to keep a good lookout for he had heared the Souix Say that they Should Shurley come to war in the Spring against us and Mandanes. ...

CLARK / 9*th* of March Satturday 1805

... Smoked a pipe (the greatest mark of friendship and attention) with the Chief [Le Borgne] .... the Manetarree Chief ... recived of Captain M. Lewis a Medel Gorget armbands, a Flag Shirt, scarlet &c ... 2 guns were fired for this Great man.

CLARK / 21st March Thursday 1805—

... on my return to day to the Fort I came on the points of the high hills,   Saw an emence quantity of Pumice Stone on the Sides & foot of the hills and emence beds of Pumice Stone near the Tops ... with evident marks of the Hills haveing once been on fire....

CHARLES MCKENZIE / [22 March 1805]

Mr. La Rocque and I [of the North West Company] ... became intimate with the gentlemen of the American expedition, who on all occasions seemed happy to see us, and always treated us with civility and kindness. It is true, Captain Lewis could not make himself agreeable to us. He could speak fluently and learnedly on all subjects, but his inveterate disposition against the British stained ... all his eloquence. Captain Clarke was equally well informed, but his conversation was always pleasant, for he seemed to dislike giving offence unnecessarily....

Fort Mandan,
1609 miles up the Missouri,
lat.47 21 47, N. long.101 25, W.
April 2d, 1805.

Dear Sir [Wm. Henry Harrison]
... The trade of the nations at this place is from the N.W. and Hudsons bay establishments on the Assinneboin river, distant about 150 miles; those traders are nearly at open war with each other, and better calculated to destroy then promote the happiness of those nations to whom they have latterly extended their trade, and intend to form an establishment near this place in the course of this year.

Your most obdt. servt.
Wm. Clark.

CLARK / April the 3rd Thursday 1805—

... we are all day engaged packing up Sundery articles to be sent to the President of the U.S.

GASS / Friday 5th April 1805

... some readers will perhaps expect ... when we are about to renew our voyage, to give some account of the fair sex of the Missouri; and entertain them with narratives of feats of love, as well as of arms. Though we could furnish a sufficient number of entertaining stories, and pleasant anecdotes, we do not think it prudent to swell our Journal with them; as our views are directed to more useful information. Besides ... we are yet ignorant of the dangers which may await us.... It may be observed generally that chastity is not very highly esteemed by these people, and that the severe and loathsome effects of certain French principles are not uncommon among them. The fact is, that the women are generally considered an article of traffic, and indulgencies are sold at a very moderate price. As a proof of this, I will just mention, that for an old tobacco-box, one of our men was granted the honour of passing a night with the daughter of the head chief of the Mandan nation. An old bawd with her punks, may also be found in some of the villages on the Missouri, as well as in the large cities of polished nations.

Fort Mandan, April 7th 1805.

Dear Sir [President Jefferson]: Herewith inclosed you will receive an invoice of ... 67. specimens of earths, salts and minerals; and 60 specimens of plants....

You will also receive herewith inclosed a part of Capt Clark's private journal.... we have encouraged our men to keep journals, and seven of them do so....

I have transmitted ... information relative to the geography of the country which we possess, together with a view of the Indian nations....

I can foresee no material or probable obstruction to our progress, and entertain therefore the most sanguine hopes of complete success.... At this moment, every individual of the party are in good health, and ... act with the most perfect harmoney. with such men I have every thing to hope, and but little to fear....

Meriwether Lewis
Capt. 1st. U'S. Regt Infty.

LEWIS / Fort Mandan April 7th. 1805.

... at 4. P.M. completed every arrangement necessary for our departure.... Our vessels consisted of six small canoes, and two large perogues. This little fleet altho' not quite so rispectable as those of Columbus or Capt. Cook, were still viewed by us with as much pleasure as those deservedly famed adventurers ever beheld theirs; and I dare say with quite as much anxiety for their safety and preservation. we were now about to penetrate a country at least two thousand miles in width, on which the foot of civilized man had never trodden.... the picture which now presented itself to me was a most pleasing one. entertaining as I do, the *most confident hope of succeeding in a voyage which had formed a darling project of mine for the last ten years, I could but esteem this moment of my departure as among the most happy of my life....*

ORDWAY / *Sunday 7th April 1805.*
... we all went on board fired the Swivel and Set off on our journey....

LEWIS / *April 8th* [1805]
... the wind blew hard against us, from the N.W. we therefore traveled very slowly. I walked on shore, and visited the black Cat, took leave of him after smoking a pipe as is their custom, and then proceeded on slowly by land about four miles.... we took dinner at this place, and then proceed on to oure encampment, which was on the N. side opposite to a high bluff....

April 8, 1805

LEWIS / *Wednesday April 17th 1805.*

... three beaver taken this morning .... the men prefer the flesh of this anamal .... I eat very heartily of the beaver myself, and think it excellent; particularly the tale, and liver ....

LEWIS / *Thursday April 25th 1805.*

... I could not discover the junction of the rivers immediately, they being concealed by the wood; however, sensible that it could not be distant .... we encamped on the bank [Glass's Bluffs] of the yellow stone river, 2 miles South of it's confluence with the Missouri.

LEWIS / *Friday April 26th 1805.*

... I proceeded down the river with one man in order to take a view of the confluence of this great river with the Missouri.... joined the party at their encampment on the point of land formed by the junction of the rivers; found them all in good health, and much pleased at having arrived at this long wished for spot, and in order to add in some measure to the general pleasure which seemed to pervade our little community, we ordered a dram to be issued to each person; this soon produced the fiddle, and they spent the evening with much hilarity, singing & dancing, and seemed as perfectly to forget their past toils, as they appeared regardless of those to come.

GASS / *Friday [April] 26th. [1805]*

... The river Jaune [Yellowstone] is shallow, and Missouri deep and rapid. In the evening Captain Lewis with his party joined us; and had brought with them a buffaloe calf, which followed them seven or eight miles. ...

WHITEHOUSE / *Thursday 2nd May 1805.*

at day light it began to Snow & blow So that we did not Set off this morning.... the men who was out a hunting found Several peaces of red cloath at an Indian camp, where we expect they left last winter for a Sacrifice to their maker as that is their form of worship, as they have Some knowledge of the Supreme being, and anything above their comprihention they call big medicine....

LEWIS / *Friday May 3rd 1805*

... the country in the neighbo of this river, and as far as the eye can reach, is level, fertile, open and beatifull beyond discription. ...

April 25, 1805

April 26, 1805

LEWIS / *Thursday May 9th 1805.*

... *Capt. Clark killed* ... *2 buffaloe, I also killed one buffaloe which proved to be the best meat* ... *we saved the best meat, and from the cow I killed we saved the necessary materials for making what our wrighthand cook Charbono, calls boudin (poudingue) blanc, and immediately set him about preparing them for supper; this white pudding we all esteem one of the greatest delicacies of the forrest* .... *About 6 feet of the lower extremity of the large gut of the Buffaloe is the first morsel* ... *the mustle lying underneath the shoulder blade next to the back, and fillets are next saught, these are needed up very fine with a portion of kidney suit; to this composition is then added a just proportion of pepper and salt and a small quantity of flour* ... *all is compleatly filled with something good to eat, it is tyed at the other end, but not any cut off, for that would make the pattern too scant; it is then baptised in the missouri with two dips and a flirt, and bobbed into the kettle; from whence, after it be well boiled it is taken and fried in bears oil until it becomes brown, when it is ready to esswage the pangs of a keen appetite, or such as travelers in the wilderness are seldom at a loss for.* ...

May 9, 1805

WHITEHOUSE / *Thursday 9th May 1805.*

clear and pleasant. we Set off at Sun rise and proceeded on   about 9 oC. we halted to take breakfast .... *the Game is getting So pleanty and tame in this country that Some of the men has went up near enofe to club them* .... *the country for Several days back is pleasant, the Soil good* ....

LEWIS / Saturday May 11th 1805.
... About 5.P.M. my attention was struck by one of the Party runing at a distance towards us and making signs and hollowing as if in distress ... I immediately turned out with seven of the party in quest of this monster [grizzly], we at length found his trale and persued him about a mile by the blood ... these bear being so hard to die reather intimedates us all; I must confess that I do not like the gentlemen and had reather fight 2 Indians than one bear; there is no other chance to conquer them by a single shot but by

May 11, 1805

shooting them through the brains .... the flece and skin were as much as two men could possibly carry. ... directed the two cooks to render the bear's oil and put it in the kegs which was done. there was about eight gallons of it.

LEWIS / Thursday May 16th [1805]
... by 4 oClock in the evening our Instruments, Medicine, merchandize provision &c, were perfectly dryed, repacked and put on board the perogue.... our medicine sustained the greatest injury .... the Indian woman to whom I ascribe equal fortitude and resolution ... caught and preserved most of the light articles which were washed overboard.

LEWIS / Friday May 17th [1805]
... Capt. Clark narrowly escaped being bitten by a rattlesnake in the course of his walk, the party killed one this evening at our encampment ....

ORDWAY / Saturday 18th May 1805.
... the Missourie is gitting clear and gravelly bottom, & Shore   we passed no falling in banks as we did below these pitch pine hilly country. a pleasant warm afternoon. ...

LEWIS / Monday May 20th 1805.
... I saw two large Owls with remarkable long feathers on the sides of the head which resembled ears ....

May 20, 1805

May 17, 1805

LEWIS / *Friday May 24th 1805.*

... the high country in which we are at present and have been passing for some days I take to be a continuation of what the Indians as well as the French Engages call the Black hills. This tract of country so called consists of a collection of high broken and irregular hills and short chain of mountains sometimes 120 miles in width and again becomeing much narrower, but always much higher than the country on either side; they commence about the head of the Kanzas river ... passing the river platte

May 24, 1805

above the forks and intercepting the Yellowstone river near the bigbend and passing the Missouri at this place and probably continuing to swell the country as far North as the Saskashawan river.... the black hills in their course nothwardly appear to approach more nearly to the Rocky Mountains....

CLARK / *May 25th Satturday 1805*

... the bottoms between hills and river are narrow and contain scercely any timber. The appeerence of salts, and bitumun still continue. we saw a polecat to day being the first which we have seen for some time past. The Air of this quarter is pure and helthy. the water of the Missouri will tasted not quite so muddy ....

May 26, 1805

May 26, 1805

LEWIS / *Sunday May 26th 1805.*

Set out at an early hour and proceeded principally by the toe line . . . . Capt. Clark walked on shore this morning and ascended to the summit of the river hills he informed me on his return that he had seen mountains . . . . In the after part of the day I also walked out and . . . thought myself well repaid for my labour; as from this point I beheld the Rocky Mountains for the first time . . . . while I viewed these mountains I felt a secret pleasure in finding myself so near the head of the heretofore conceived boundless Missouri; but when I reflected on the difficulties which this snowey barrier would most probably throw in my way to the Pacific, and the sufferings and hardships of myself and party in thim, it in some measure counterballanced the joy I had felt in the first moments in which I gazed on them; but as I have always held it a crime to anticipate evils I will believe it a good comfortable road untill I am compelled to believe differently. . . . on my return to camp I trod within few inches of a rattle snake but being in motion I . . . fortunately escaped his bite, I struck about with my espontoon being directed in some measure by his nois untill I killed him. . . . The appearances of coal in the face of the bluffs, also of birnt hills, pumice stone salts and quarts continue as yesterday. This is truly a desert barren country and I feel myself still more convinced of it's being a continuation of the black hills. . . .

CLARK / *May 26th Sunday 1805*

. . . I crossed a Deep holler and assended a part of the plain elivated much higher . . . from this point I beheld the Rocky Mountains for the first time with certainty . . . . those points of the rocky Mountain were covered with Snow and the Sun Shown on it in such a manner as to give me a most plain and satisfactory view. . . .

May 26, 1805

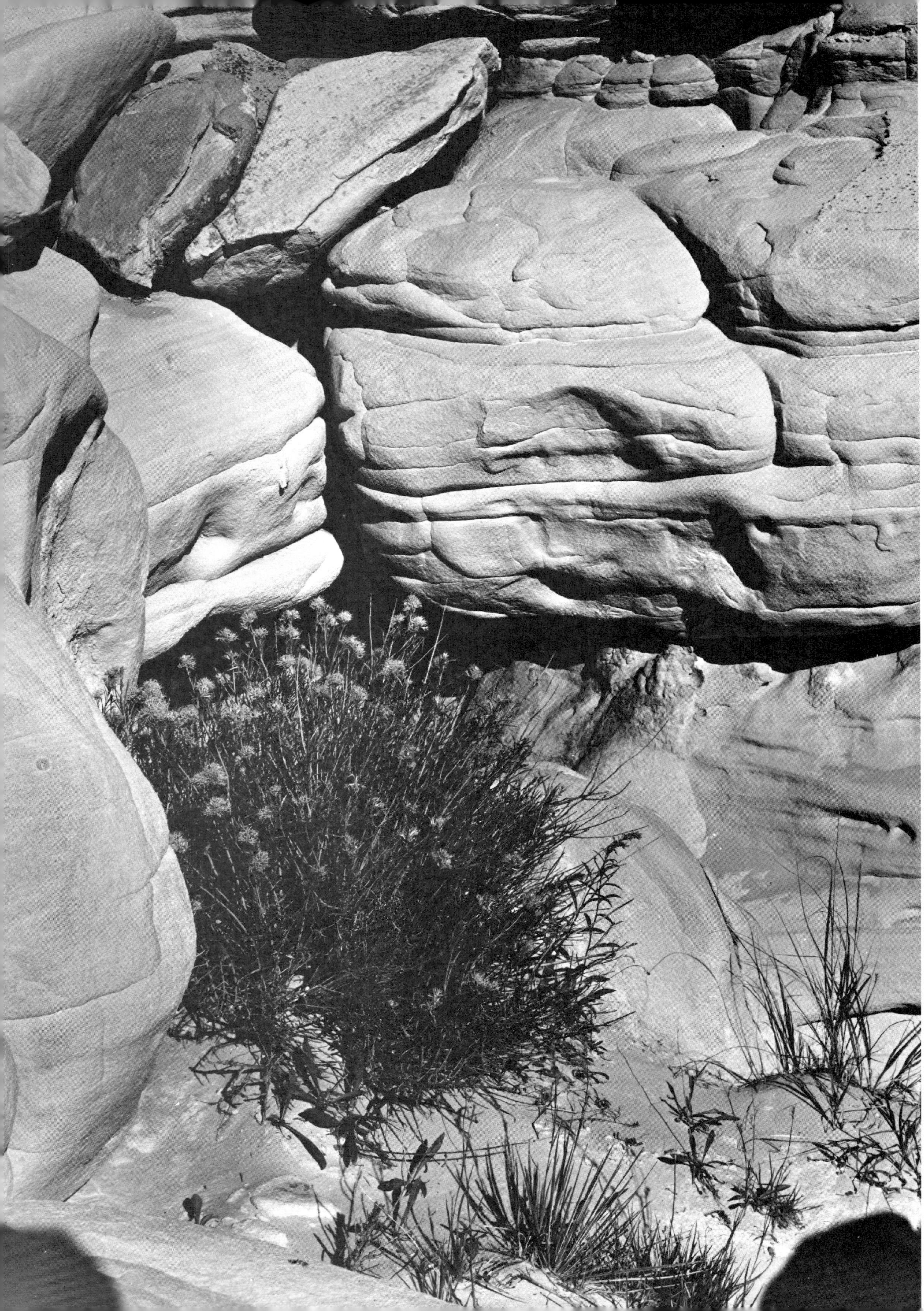

May 27, 1805

LEWIS / Monday May 27.*th* 1805.

... the bluffs are very high steep rugged, containing considerable quantities of stone and border the river closely on both sides; once perhaps in the course of several miles there will be a few acres of tolerably level land in which two or thre impoverished cottonwood trees will be seen....

LEWIS / *Monday May 27th 1805.*
... the bluffs are composed of irregular tho' horizontal stratas of yellow and brown or black clay, brown and yellowish white sand, of soft yellowish white sandstone and a hard dark brown freestone, also ... irregular seperate masses of a hard black Iron stone, which is imbeded in the Clay and sand. some little pine spruce and dwarf cedar on the hills. some coal or carbonated wood still makes it's appearance in these bluffs .... the country more broken and barren than yesterday if possible. about midday it was very warm to this the high bluffs and narrow channel of the river no doubt contributed greatly. ...

GASS / *Monday [May] 27th [1805]*
... the most dismal country I ever beheld. ...

May 27, 1805

May 27, 1805

May 28, 1805

LEWIS / *Tuesday May 28th 1805.*

... the river spreads to more than 3 times it's former width and is filled with a number of small and handsome Islands covered with cottonwood   some timber ... the land again fertile. these appearances were quite reviving after the drairy country through which we had been passing. . . . we encamped on Stard opposite to the entrance of a small Creek [Dog Creek]. . . .

LEWIS / *Wednesday May 29th 1805*

Last night we were all allarmed by a large buffaloe Bull, which swam over from the opposite shore and coming along side of the white perogue, climbed over it to land, he then allarmed ran up the bank in full speed directly towards the fires, and was within 18 inches of the heads of some of the men who lay sleeping before the centinel could allarm him or make him change his course, still more alarmed, he now took his direction immediately towards our lodge, passing between 4 fires and within a few inches of the heads of one range of the men as they yet lay sleeping, when he came near the tent, my dog saved us by causing him to change his course a second time, which he did by turning a little to the right, and was quickly out of sight, leaving us by this time all in an uproar with our guns in our hands, enquiring of each other the cause of the alarm, which after a few moments was explained by the centinel: we were happy to find no one hirt. . . .

May 29, 1805

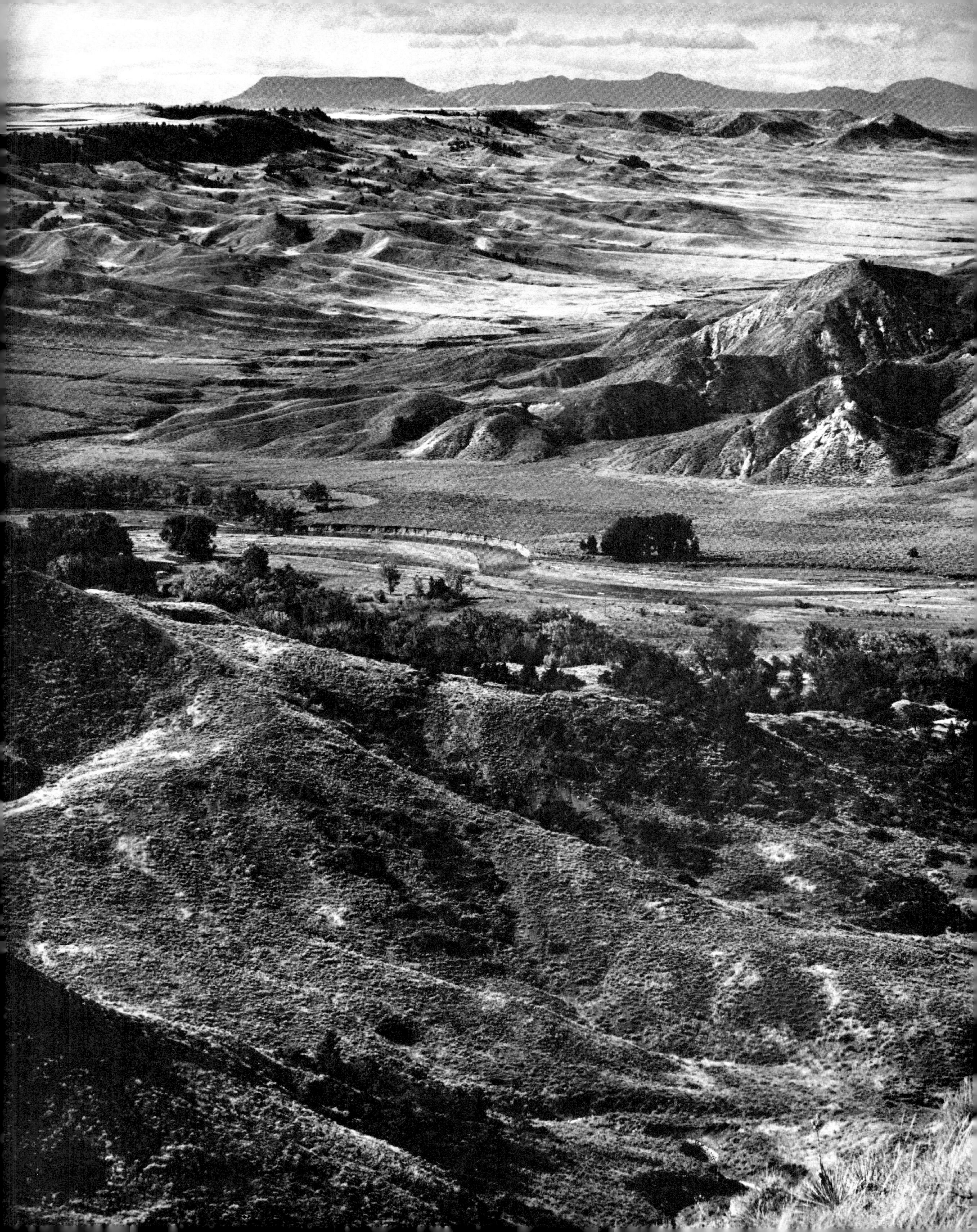

May 29, 1805

Lewis / *Wednesday May 29th 1805*

... I walked on shore and acended this river about a mile and a half in order to examine it.... the bed was formed of gravel and mud with some sand ... it was more rappid but equally navigable; there were no large stone or rocks in it's bed to obstruct the navigation; the banks were low yet appeared seldom to overflow; the water of this River is clearer much than any we have met with   great abundance of the Argalia or Bighorned animals in the high country through which this river passes. Cap. C. who assended this R. much higher than I did has thought proper to call (called) it Judieths [Big Horn] River. the bottoms of this stream as far as I could see were wider and contained more timber than the Missouri; here I saw some box alder intermixed with the Cottonwood willow; rose bushes and honeysuckle with some red willow constitute the undergrowth.

Clark / *May 29th Wednesday 1805*

... we came too for Dinner opposit the enterence of a small river .... and as we were in a good harbor & small point of woods on the Stard Side, and no timber for some distance above, induced us to conclude to stay all night. we gave the men a dram, altho verry small it was sufficent to effect several men....

May 29, 1805

May 29, 1805

May 30, 1805

LEWIS / Thursday May 30th 1805.
... many circumstances indicate our near approach to a country whos climate differs considerably from that in which we have been for many months. the air of the open country is asstonishingly dry as well as pure.... I also observed the well seasoned case of my sextant shrunk considerably and the joints opened. The water of the river still continues to become clearer ... than it was a few days past. this day we proceded with more labour and difficulty than we have yet experienced ... the banks and sides of the bluff were more steep than usual and were now rendered so slippery by the late rain that the men could scarcely walk.... the earth and stone also falling from these immence high bluffs render it dangerous to pass under them. the wind was also hard and against us. our chords broke several times today but happily without injury to the vessels. ...

May 30, 1805

CLARK / *May 30th Thursday 1805*

... We discover in several places old encampments of large bands of Indians ... we believe to be the Blackfoot Inds or Menitares who inhabit the heads of the Saskashowin & north of this place. and trade a little in the Fort de Prarie [Edmonton, Alta.] *establishments. we camped in a grove of cotton trees on the Stard Side, river rise 1½ In.*

LEWIS / *Friday May 31st 1805.—*

... The toe rope of the white perogue ... gave way today at a bad point ... was very near overseting; I fear her evil gennii will play so many pranks with her that she will go to the bottomm some of those days. ...

The hills and river Clifts which we passed today exhibit a most romantic appearance. The bluffs of the river rise to the hight of from 2 to 300 feet and in most places nearly perpendicular; they are formed of remarkable white sandstone ... two or thre thin horizontal stratas of white freestone, on which the rains or water make no impression, lie imbeded in these clifts of soft stone near the upper part of them ...

May 30, 1805

May 31, 1805

May 31, 1805

May 31, 1805

May 31, 1805

May 31, 1805

LEWIS / *Friday May 31st 1805.—*
... the earth on the top of these Clifts is a dark rich loam, which forming a graduly ascending plain extends back from ½ a mile to a mile where the hills commence and rise abruptly to a hight of about 300 feet more. The water in the course of time in decending from those hills and plains on either side of the river has trickled down the soft sand clifts and woarn it into a thousand grotesque figures, which with the help of a little immagination and an oblique view, at a distance are made to represent eligant ranges of lofty freestone buildings, having their parapets well stocked with statuary ....

May 31, 1805

May 31, 1805

LEWIS / *Friday May 31ˢᵗ 1805.—*
... collumns of various sculpture both grooved and plain, are also seen supporting long galleries in front of those buildings; in other places on a much nearer approach and with the help of less immagination we see the remains or ruins of eligant buildings; some collumns standing and almost entire with their pedestals and capitals; others retaining their pedestals but deprived by time or accident of their capitals, some lying prostrate an broken others in the form of vast pyramids of connic structure bearing a serees of other pyramids on their tops becoming less as they ascend and finally terminating in a sharp point. nitches and alcoves of various forms and sizes are seen at different hights as we pass. a number of the small martin which build their nests with clay in a globular form attatched to the wall within those nitches, and which were seen hovering about the tops of the collumns did not the less remind us of some of those large stone buildings in the U. States. ... the thin stratas of hard freestone intermixed with the soft sandstone seems to have aided the water in forming this curious scenery. As we passed on it seemed as if those seens of visionary inchantment would never have [an] end; for here it is too that nature presents to the view of the traveler vast ranges of walls of tolerable workmanship .... these walls sometimes run parallel to each other, with several ranges near each other, and at other times intersecting each other at right angles, having the appearance of the walls of ancient houses or gardens. ...

May 31, 1805

LEWIS / *Friday May 31st 1805.—*

... I walked on shore this evening and examined these walls minutely and preserved a specimine of the stone. . . . on these clifts I met with a species of pine which I had never seen . . . .

CLARK / *May 31st Friday 1805*

. . . Capt Lewis . . . collected some of the stones off one of the walls which appears to be a sement of Isin glass [and] black earth   we camped on the Stard Side in a small timbered bottom above the mouth of a Creek on the Stard Side . . . .

LEWIS / *Sunday June 2nd 1805.*

. . . The river bluffs still continue to get lower and the plains leveler and more extensive; the timber on the river increases in quantity . . . . killed 6 Elk 2 buffale 2 Mule deer and a bear . . . . the bear was very near catching Drewyer; it also pursued Charbono who fired his gun in the air as he ran but fortunately eluded the vigilence of the bear by secreting himself very securely in the bushes untill Drewyer finally killed it . . . .

May 31, 1805

May 31, 1805

LEWIS / *Monday June 3rd 1805.*

... An interesting question was now to be determined; which of these rivers was the Missouri, or that river which the Minnetares call Amahte Arzzha or Missouri, and which they had discribed to us as approaching very near to the Columbia river. to mistake the stream ... would not only loose us the whole of this season but would probably so dishearten the party that it might defeat the expedition altogether .... to this end an investigation of both streams was the first thing to be done .... thus have our cogitating faculties been busily employed all day. ...

ORDWAY / *June 6th Thursday 1805.*

... about 2 oClock P.M. Capt Clark and his party returned to Camp had been about 40 miles up the South fork & Capt Clark thinks it will be the best course for us to go. ...

LEWIS / *Saturday June 8th 1805.—*

... The whole of my party to a man except myself were fully pesuaided that this river was the Missouri, but being fully of opinion that it was neither the main stream, nor that which it would be advisable for us to take, I determined to give it a name and in honour of Miss Maria W————d. called it Maria's River. it is true that the hue of the waters of this turbulent and troubled stream but illy comport with the pure celestial virtues and amiable qualifications of that lovely fair one; but on the other hand it is a noble river; one destined to become in my opinion an object of contention between the two great powers of America and Great Britin with rispect to the adjustment of the Northwestwardly boundary of the former; and that it will become one of the most interesting brances ... in a commercial point of view ... as it abounds with anamals of the fur kind, and most probably furnishes a safe and direct communication to that productive country of valuable furs exclusively enjoyed at present by the subjects of his Britanic Majesty ....

ORDWAY / *Saturday 8th June 1805.*

... So our Captains conclude to assend the South South fork .... they named the North fork River Mariah and the middle or little River named Tanzey [Teton] River. The water & bottoms in everry respect of each resimbles the Missourie below the forks. only Smaller. ...

WHITEHOUSE / *Sunday 9th June 1805.*

... towards evening we had a frolick. the officers gave the party a dram, the fiddle played and they danced late &. ...

LEWIS / *Thursday June 13th 1805.*

... from the extremity of this roling country I overlooked a most beatifull and level plain of great extent or at least 50 or sixty miles .... curious mountains presented themselves of square figures, the sides rising perpendicularly to the hight of 250 feet and appeared to be formed of yellow clay; their tops appeared to be level plains .... I did not however loose my direction to this point which soon began to make a roaring too tremendious to be mistaken for any cause short of the great falls of the Missouri. here I arrived about 12 OClock ....

June 8, 1805

LEWIS / *Thursday June 13th 1805.*

... from the reflection of the sun on the sprey or mist which arrises from these falls there is a beatifull rainbow produced which adds not a little to the beauty of this majestically grand senery. after wrighting this imperfect discription I again viewed the falls and was so much disgusted with the imperfect idea which it conveyed of the scene that I determined to draw my pen across it and begin agin, but then reflected that I could not perhaps succeed better than pening the first impressions of the mind; I wished for the pencil of Salvator Rosa [a Titian] or the pen of Thompson, that I might be enabled to give to the enlightened world some just idea of this truly magnifficent and sublimely grand object, which has from the commencement of time been concealed from the view of civilized man; but this was fruitless and vain. I most sincerely regreted that I had not brought a crimee [camera] obscura with me by the assistance of which even I could have hoped to have done better but alas this was also out of my reach ....
... My fare is really sumptuous this evening; buffaloe's humps, tongues and marrowbones, fine trout parched meal pepper and salt, and a good appetite; the last is not considered the least of the luxuries.

LEWIS / *Friday June 14th 1805.*

... hearing a tremendious roaring above me I continued my rout across the point of a hill a few hundred yards further and was again presented by one of the most beatifull objects in nature, a cascade of about fifty feet perpendicular streching at rightangles across the river from side to side to the distance of at least a quarter of a mile.... I now thought that if a skillfull painter had been asked to make a beautifull cascade that he would most probably have pesented the precise immage of this one; nor could I for some time determine on which of those two great cataracts to bestoe the palm, on this or that which I had discovered yesterday; at length I determined between these two great rivals for glory that this was pleasingly beautifull, *while the other was* sublimely grand.

CLARK / *June 14th Friday 1805*

... Jo: Fields returned from Capt Lewis with a letter for me, Capt Lewis ... is convinced of this being the river the Indians call the Missouri ....

June 14, 1805

June 14, 1805

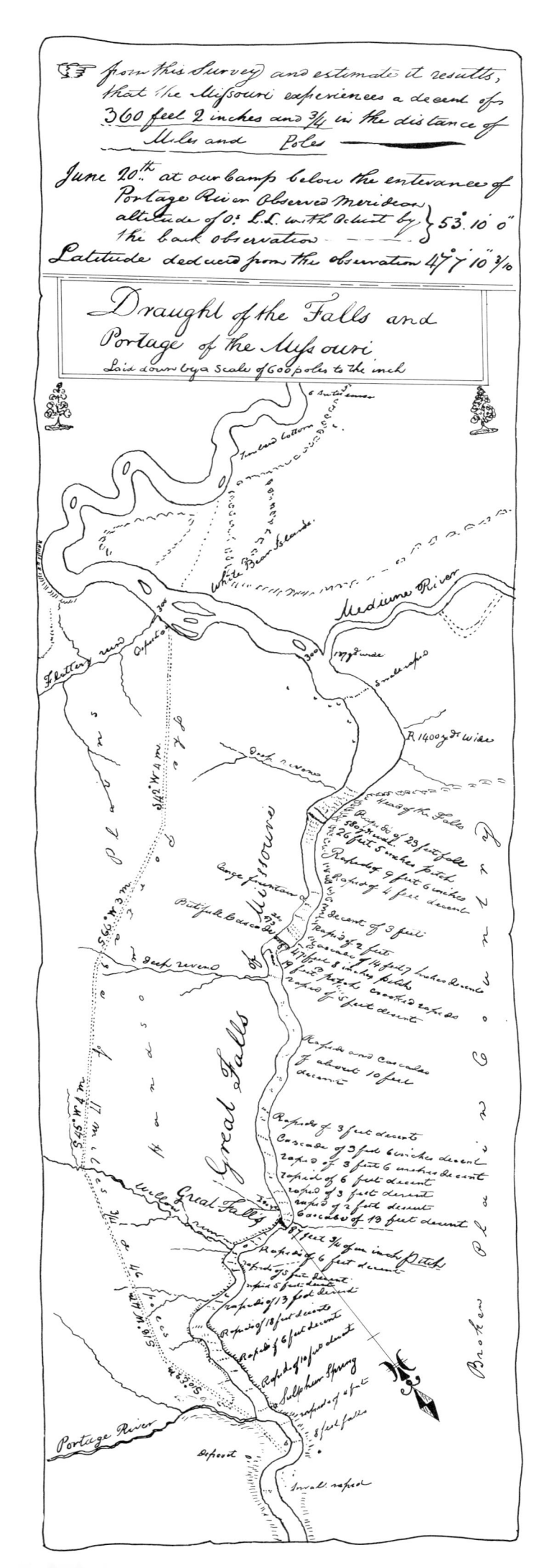

LEWIS / Sunday June 16th 1805.

.... about 2 P.M. I reached the camp found the Indian woman [Sacajawea] extreemly ill and much reduced by her indisposition. this gave me some concern as well for the poor object herself, then with a young child in her arms, as from the consideration of her being our only dependence for a friendly negociation with the Snake Indians on whom we depend for horses to assist us in our portage from the Missouri to the columbia river. I now informed Cap.t C. of my discoveries with rispect for our portage . . . which I could not estimate at less than 16 miles. . . .

LEWIS / Monday June 17th 1805.

... I set six men at work to prepare four sets of truck wheels with couplings, toungs and bodies, that they might either be used without the bodies for transporting our canoes, or with them in transporting our baggage ....

ORDWAY / Monday 17th June 1805.

... 2 hunters out in order to git Elk Skins to cover or bottom our Iron boat when we git ab.o the falls, as we will Stand in need of it, as we leave our largest craft at this place. . . .

LEWIS / Tuesday June 18th 1805.

This morning I employed all hands in drawing the perogue on shore in a thick of willow bushes .... I now scelected a place for a cash [cache] and set three men at work to complete it, and employed all others except those about the waggons, in overhawling airing and repacking our indian goods ammunicion, provision and stores of every discription which required inspection. examined the frame of my Iron boat and found all parts complete except one screw, which the ingenuity of Sheilds can readily replace, a resource which we have very frequent occasion for. . . .

LEWIS / Wednesday June 19th 1805.

... the Indian woman was much better this morning she walked out and gathered a considerable quantity of the white apples of which she eat so heartily in their raw state, together with a considerable quantity of dryed fish without my knowledge that she complained very much and her fever again returned. I rebuked Sharbono severely for suffering her to indulge herself with such food .... I now gave her broken dozes of diluted nitre untill it produced perspiration and at 10 P.M. 30 drops of laudnum which gave her a tolerable nights rest. . . .

LEWIS / *Saturday June 22nd 1805.*

This morning early Cap.t Clark and myself with all the party except Serg.t Ordway Sharbono, Goodrich, York and the Indian woman, set out to pass the portage with the canoe and baggage to the Whitebear Island, where we intend that this portage shall end.   Cap.t Clark piloted us through the plains. . . .

LEWIS / *Sunday June 23rd 1805.*

. . . this evening the men repaired their mockersons, and put on double souls to protect their feet from the prickley pears. . . . they are obliged to halt and rest frequently for a few minutes, at every halt these poor fellows tumble down and are so much fortiegued that many of them are asleep in an instant; in short their fatiegues are incredible; some are limping from the soreness of their feet, others faint and unable to stand for a few minutes, with heat and fatiegue, yet no one complains, all go with cheerfullness. . . .

WHITEHOUSE / *Monday 24th June 1805.*

a fair morning. we halled out the last canoe . . . . Set out eairly with a waggon & baggage . . . . had Some difficulty in gitting the loading up on the high plains . . . . we hoisted a Sail in the largest canoe which helped us much as 4 men halling at the chord with a harness. passed through high Smoth delightful plains. . . .

LEWIS / *Tuesday June 25th 1805.*

. . . The party that returned this evening to the lower camp reached it in time to take one canoe on the plain and prepare their baggage for an early start in the morning after which such as were able to shake a foot amused themselves in dancing on the green to the music of the violin which Cruzatte plays extreemly well. . . .

LEWIS / *Wednesday June 26th 1805.*

The Musquetoes are extreemly troublesome to us. . . . set [Pvt. Robert] Frazier at work to sew the skins together for the covering of the boat. . . . and to myself I assign the duty of cook . . . . I collected my wood and water, boiled a large quantity of excellent dryed buffaloe meat and made each man a large suet dumpling by way of a treat. . . .

LEWIS / *Saturday June 29th 1805.*

. . . I have scarcely experienced a day since my first arrival in this quarter without experiencing some novel occurrence among the party or witnessing the appearance of some uncommon object. . . . about 25 y.ds from the river . . . . the water of the fountain boil up with such force near it's center that it's surface in that part seems even higher than the surrounding earth which is a firm handsom terf of fine green grass. after amusing myself about 20 minutes in examining the fountain I found myself so chilled with my wet cloaths that I determined to return and accordingly set out . . . .

June 29, 1805

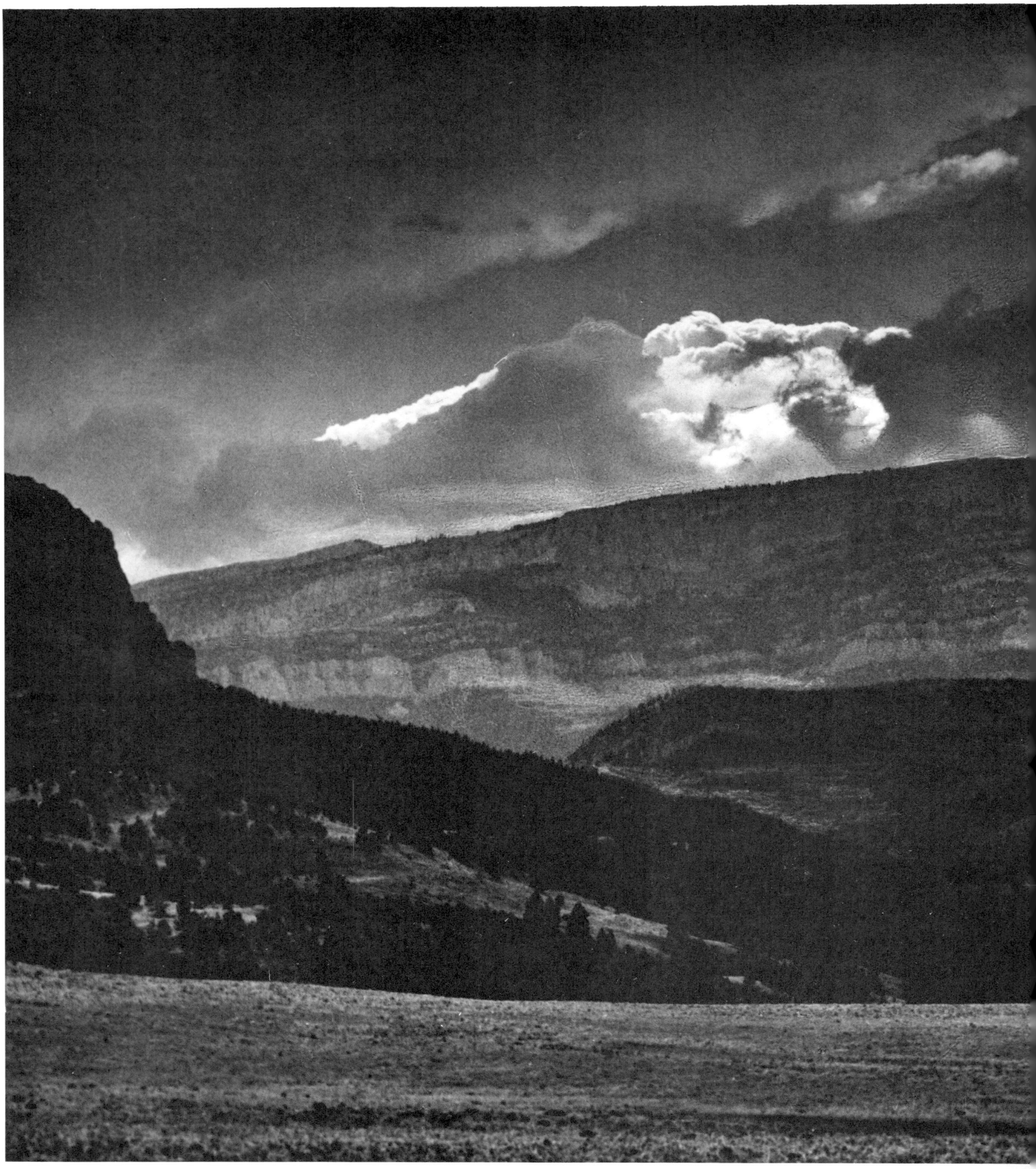

WHITEHOUSE / June 29th Saturday 1805.

... in the afternoon their arose a storm of hard wind and rain and amazeing large hail at our Camp we measured & weighed Some of them, and Cap‡ Lewis made a bowl of Ice punch of one of them   they were 7 Inches in Surcumference and weighed 3 ounces....

ORDWAY / June 29th Satturday 1805.

... Saw a black cloud rise in the west .... ran in great confusion to Camp the hail being so large and the wind so high and violent in the plains, and we being naked we were much bruuzed by the large hail. Some nearly killed   one knocked down three times ....

CLARK / June 29th Satturday 1805

... the rain fell like one voley of water falling from the heavens and gave us time only to get out of the way of a torrent of water which was Poreing down the hill in [to] the River with emence force tareing every thing before it takeing with it large rocks & mud, I took my gun & shot pouch in my left hand, and with the right scrambled up the hill pushing the Interpreters wife (who had her child in her arms) before me, the Interpreter himself makeing attempts to pull up his wife by the hand much scared and nearly without motion, we at length reached the top of the hill .... before I got out of the bottom of the reveen which was a flat dry rock when I entered it, the water was up to my waste & wet my watch, I scercely got out before it raised 10 feet deep with a torrent which [was] turrouble to behold .... I lost at the river in the torrent the large compas, an elegant fusee, Tomahawk Humbrallo ... &ᶜ ...

CLARK / June 30th Sunday 1805

... Men complain of being Soore this day dull and lolling about .... Great numbers of Buffalow in every derection. I think about 10,000 may be seen in a view.

June 29, 1805

July 2, 1805

Clark / July 1st Monday 1805.

We set out early this morning with the remaining load, and proceeded on verry well to Capt Lewis's camp where we arrived at 3 oClock, the Day worm and party much fatigued, found Capt Lewis and party all buisey employed in fitting up the Iron boat . . . . The hunters killed 3 white bear [grizzlies] one large, the fore feet of which measured 9 inches across, the hind feet 11 Inches ¾ long & 7 Inches wide a bear [came] naarly catching Joseph Fields chased him into the water, bear about the camp every night and seen on an [White Bear] Isl^d in the day

Lewis / Tuesday July 2nd 1805.

. . . about 2 P.M. the party returned with the baggage, all well pleased that they had completed the laborious task of portage. . . . After I had completed my observation of Equal altitudes today Capt Clark Myself and 12 men passed over to the large Island to hunt bear. . . . this brush we entered in small parties of 3 or four together and surched in every part. we found one only which made at Drewyer and he shot him in the brest at the distance of about 20 feet, the ball fortunately passed through his heart, the stroke knocked the bear down and gave Drewyer time to get out of his sight; the bear changed his course we pursued him about 100 yards by the blood and found him dead . . . . after our return, in moving some of the baggage we caught a large rat. (Copy for Dr Barton) it was somewhat larger than the common European rat, of lighter colour . . . . the whiskers very long and full. the tail was . . . covered with fine fur or poil of the same length and colour of the back. the fur was very silkey close and short. . . .

Lewis / Wednesday July 3rd 1805.

. . . Indians have informed us that we should shortly leave the buffaloe country after passing the falls; this I much regret for . . . we shal sometimes be under the necessity of fasting occasionally. and at all events the white puddings will be irretrievably lost and Sharbono out of imployment. . . . The current of the river looks so gentle and inviting that the men all seem anxious to be moving upwards as well as ourselves. . . .

Lewis / Thursday July 4th 1805.

. . . have concluded not to dispatch a canoe with a part of our men to St Louis as we had intended early in the spring. we fear also that such a measure might possibly

July 7, 1805

discourage those who would in such case remain, and might possibly hazzard the fate of the expedition. we have never once hinted to any one of the party that we had such a scheme in contemplation, and all appear perfectly to have made up their minds to suceed in the expedition or purish in the attempt. we all beleive that we are now about to enter on the most perilous and difficult part of our voyage, yet I see no one repining; all appear ready to me[e]t those difficulties which await us with resolution and becoming fortitude. . . . the Mountains to the N.W. & W. of us . . . might have derived their appellation of shining Mountains, from their glittering appearance when the sun shines in certain directions on the snow which covers them. . . .

ORDWAY / *July 4th Thursday 1805.*
. . . it being the 4th of Independence we drank the last of our ardent Spirits except a little reserved for Sickness. the fiddle put in order and the party amused themselves dancing all the evening untill about 10 oClock in a Sivel & jovil manner. . . .

LEWIS / *Sunday July 7th 1805.*
The weather warm and cloudy . . . . we dispatched two other hunters to kill Elk or buffaloe for their skins to cover our baggage. . . . Capt Clarks black man York is very unwell today and he gave him a doze of tartar emettic which operated very well and he was much better in the evening. . . .

WHITEHOUSE / *July 8th Monday 1805.*
. . . about 9 oClock A. M. Capt Clark and all the men that could be Spared from Camp Set out for to go down to the falls a hunting. I remained in Camp making leather cloathes &c.   the rest of the men at Camp was employed in makeing coal & tallow and Beese wax mixed and payed over the leather on the Iron boat &c. . . .

LEWIS / *Tuesday July 9th 1805.*
. . . launched the boat [the experiment]; she lay like a perfect cork in the water. five men would carry her with the greatest ease. . . . late in the evening . . . she leaked in such manner that she would not answer. I need not add that this circumstance mortifyed me not a little . . . the evil was irraparable . . . . had I only singed my Elk skins in stead of shaving them I beleive the composition [caulking] would have remained and the boat have answered . . . . but it was now too late to introduce a remidy and I bid adieu to my boat, and her expected services. . . .

ORDWAY / *July 9th Tuesday 1805.*
. . . So we Sank hir in the water So that She might be the easier took to peaces tomorrow. our officers conclude to build 2 canoes more So that we can carry all our baggage without the Iron boat. about 10 men got ready to up the river to build 2 canoes.

July 13, 1805

July 15, 1805

Lewis / Saturday July 13th 1805.

... from the head of the white bear Islands I passed in a S.W. direction and struck the Missouri at 3 Miles and continued up it to Capt Clark's camp where I arrived about 9 A.M. and found them busily engaged with their canoes Meat &c. ...
... the hunters killed three buffaloe today which were in good order. the flesh was brought in dryed the skins wer also streached for covering our baggage. we eat an emensity of meat; it requires 4 deer, an Elk and a deer, or one buffaloe, to supply us plentifully 24 hours. meat now forms our food prinsipally as we reserve our flour parched meal and corn as much as possible for the rocky mountains which we are shortly to enter .... The Musquetoes and knats are more troublesome here if possible ....

Lewis / Monday July 15th 1805

We arrose very early this morning, assigned the canoes their loads and had it put on board.... we find it extreemly difficult to keep the baggage of many of our men within reasonable bounds; they will be adding bulky articles of but little uce or value to them. At 10 A.M. we once more saw ourselves fairly under way much to my joy and I beleive that of every individual who compose the party.... Drewyer wounded a deer which ran into the river my dog pursued caught it drowned it and brought it to shore at our camp. we have now passed Fort Mountain [Square Butte] on our right it appears to be about ten miles distant. ... it's sides stand nearly at right angles with each other and are each about a mile in extent. these are formed of a yellow clay ... and rise perpendicularly to the hight of 300 feet. the top appears to be a level plain and ... covered with a similar coat of grass with the plain on which it stands. ...

July 16, 1805

LEWIS / *Tuesday July 16th 1805.*

... Drewyer killed a buffaloe this morning ... we halted and breakfasted on it. here for the first time I ate of the small guts of the buffaloe cooked over a blazing fire in the Indian stile without any preperation of washing or other clensing and found them very good. After breakfast I determined to leave Capt C. and party, and go on to the point where the river enters the Rocky Mountains ....

July 19, 1805

LEWIS / *Thursday July 18th 1805.*

... we were anxious now to meet with the Sosonees or snake Indians as soon as possible in order to obtain information relative to the geography of the country and also if necessary, some horses . . . .

LEWIS / *Friday July 19th 1805.*

... this evening we entered much the most remarkable clifts that we have yet seen. these clifts rise from the waters edge on either side perpendicularly to the hight of (about) 1200 feet. every object here wears a dark and gloomy aspect. the tow[er]ing and projecting rocks in many places seem ready to tumble on us. the river appears to have forced it's way through this immence body of solid rock for the distance of 5¾ Miles and where it makes it's exit below has th[r]own on either side vast collumns of rocks mountains high. . . . nor is ther in the 1st 3 Miles of this distance a spot except one of a few yards in extent on which a man could rest the soal of his foot. . . . it was late in the evening before I entered this place and was obliged to continue my rout untill sometime after dark before I found a place sufficiently large to encamp my small party; at length such an one occurred on the lard side where we found plenty of lightwood and pi[t]ch pine. . . . from the singular appearance of this place I called it the gates of the rocky mounatains. . . .

CLARK / *July 19th Fryday 1805*

... Killd two [elk] and dined being oblige[d] to substitute dry buffalow dung in place of wood .... my feet is verry much brused & cut walking over the flint, & constantly stuck full [of] Prickley pear thorns, I puled out 17 by the light of the fire to night ....

ORDWAY / *July 22nd Monday 1805.*

... Capt Lewis forgot his Thurmometer where we dined   I went back for it. it Stood in the heat of the day at 80 degrees abº 0 .... our Intrepters wife tells us that She knows the country along the River up to hir nation, or the 3 forks. we are now 166 miles from the falls of the M.   Came 17 miles of it to day.

July 19, 1805

July 19, 1805

LEWIS / *Wednesday July 24th 1805.*
... our trio of pests still invade and obstruct us on all occasions, these are the Musquetoes eye knats and prickley pears, equal to any three curses that ever poor Egypt laiboured under, except the Mahometant yoke. the men complain of being much fortiegued. their labour is excessively great. I occasionly encourage them by assisting ... and have learned to push a tolerable good pole in their fraize....

July 24, 1805

July 25, 1805

July 27, 1805

July 27, 1805

LEWIS / *Thursday July 25th 1805.*

... *we killed a couple of young gees which are very abundant and fine; but as they are but small game to subsist a party on of our strength I have forbid the men shooting at them as it waists a considerable quantity of amunition and delays our progress....*

LEWIS / *Saturday July 27th 1805.—*

at the distance of 3¾ Ms. further we arrived at 9. A.M. at the junction of the S.E. fork of the Missouri and the country opens suddonly to extensive and bea[u]tifull plains and meadows which appear to be surrounded in every direction with distant and lofty mountains; supposing this to be the three forks of the Missouri I halted the party on the Lard shore for breakfast. and walked up the S.E. fork about ½ a mile and ascended the point of a high limestone clift from whence I commanded a most perfect view of the neighbouring country.... believing this to be an essential point in geography of this western part of the Continent I determined to remain at all events untill I obtained the necessary data for fixing it's latitude Longitude &c.

LEWIS / *Sunday July 28th 1805.*

... *In pursuance of this resolution we called the S.W. fork, that which we meant to ascend, Jefferson's River in honor of that illustrious personage Thomas Jefferson. [the author of our enterprize.] the Middle fork we called Madison's River in honor of James Madison, and the S.E. Fork we called Gallitin's River in honor of Albert Gallitin. the two first are 90 yards wide and the last is 70 yards. all of them run with great volocity and th[r]ow out large bodies of water.* ...

WHITEHOUSE / *Sunday 28th July 1805*

... *Capt Clark verry unwell. we built a bowrey for his comfort. the party in general much fatigued. Several lame.... I am employed makeing the chief part of the cloathing for the party....*

LEWIS / *Tuesday July 30th 1805.*

Capt Clark being much better this morning . . . we reloaded our canoes and set out, ascending Jeffersons river. . . .

WHITEHOUSE / *Thursday 1st day of August 1805*

. . . it being Capt Clarks buthday he ordered Some flour gave out to the Party. . . .

LEWIS / *Thursday August 8th 1805.*

. . . the Indian woman recognized the point of a high plain to our right . . . . this hill she says her nation calls the beaver's head . . . . it is now all important with us to meet with those people as soon as possible I determined to proceed tomorrow with a small party to the source of the principal stream of this river and pass the mountains to the Columbia; and down that river untill I found the Indians . . . for without horses we shall be obliged to leave a great part of our stores . . . already sufficiently small for the length of the voyage before us.

ORDWAY / *Friday 9th August 1805.*

. . . Capt Lewis, Shields, Drewyer & Mcneal set out to go on a head a long distance to make discoveries in hopes to find Indians &.C. . . . we came 18 mls . . . .

LEWIS / *Saturday August 10th 1805.*

. . . from the number of rattle snakes about the Clifts at which we halted we called them the rattle snake clifts. . . . I do not beleive that the world can furnish an example of a river runing to the extent which the Missouri and Jefferson's rivers do through such a mountainous country and at the same time so navigable as they are. if the Columbia furnishes us such another example, a communication across the continent by water will be practicable and safe. but this I can scarcely hope . . . .

CLARK / *August 10th Saiturday 1805*

Some rain this morning at Sun rise and Cloudy we proceeded on passed a remarkable Clift   point on the Stard Side about 150 feet high, this Clift the Indians Call the Beavers head, opposit at 300 yards is a low clift of 50 feet which is a Spur from the Mountain on the Lard about 4 miles, the river verry Crooked . . . .

August 8, 1805

August 10, 1805

August 10, 1805

August 12, 1805

LEWIS / *Monday August 12th 1805*

... we halted and breakfasted on the last of our venison .... after eating we continued our rout through the low bottom of the main stream along the foot of the mountains .... the road was still plain, I therefore did not dispair of shortly finding a passage over the mountains and of taisting the waters of the great Columbia this evening.... at the distance of 4 miles further the road took us to the most distant fountain of the waters of the Mighty Missouri in surch of which we have spent so many toilsome days and wristless nights. thus far I had accomplished one of those great objects on which my mind has been unalterably fixed for many years, judge then of the pleasure I felt in all[a]ying my thirst with this pure and ice-cold water which issues from the base of a low mountain or hill of a gentle ascent for ½ a mile. the mountains are high on either hand leave this gap at the head of this rivulet through which the road passes. here I halted a few minutes and rested myself. two miles below M*c*Neal had exultingly stood with a foot on each side of this little rivulet and thanked his god that he had lived to bestride the mighty & heretofore deemed endless Missouri. after refreshing ourselves we proceeded on to the top of the dividing ridge from which I discovered immence ranges of high mountains still to the West of us with their tops partially covered with snow. I now decended the mountain about ¾ of a mile which I found much steeper than on the opposite side, to a handsome bold runing Creek of cold Clear water. here I first tasted the water of the great Columbia river.

August 12, 1805

August 13, 1805

LEWIS / *Tuesday August 13th 1805.*
... we set out, still pursuing the road down the river. we had marched about 2 miles when we met a party of about 60 warriors mounted on excellent horses who came in nearly full speed, when they arrived I advanced towards them with the flag leaving my gun with the party about 50 paces behind me. the chief and two others who were a little in advance of the main body spoke to the women, and they informed them who we were and exultingly shewed the presents which had been given them   these men then advanced and embraced me very affectionately in their way which is by puting their left arm over your wright sholder clasping your back, while they apply their left cheek to yours and frequently vociforate the word âh-hí-e, âh-hí-e that is, I am much pleased, I am much rejoiced. bothe parties now advanced and we wer all carressed and besmeared with their grease and paint till I was heartily tired of the national hug. I now had the pipe lit and gave them smoke; they seated themselves in a circle around us and pulled off their mockersons before they would receive or smoke the pipe. this is a custom among them as I afterwards learned indicative of a sacred obligation of sincerity in their profession of friendship .... as much as to say that they wish they may always go bearfoot if they are not sincere; a pretty heavy penalty if they are to march through the plains of their country.... the principal chief Ca-me-âh-wait made a short speach to the warriors. I gave him the flag which I informed him was an emblem of peace among whitemen and now that it had been received by him it was to be respected as the bond of union between us....

LEWIS / *Wednesday August 14th [1805]*
... The means I had of communicating with these people was by way of Drewyer who understood perfectly the common language of jesticulation or signs which seems to be universally understood by all the Nations we have yet seen. it is true that this language is imperfect and liable to error but is much less so than would be expected. the strong parts of the ideas are seldom mistaken.
... Drewyer who had had a good view of their horses estimated them at 400. most of them are fine horses. indeed many of them would make a figure on the South side of James River or the land of fine horses. I saw several with spanish brands on them, and some

mules which they informed me that they had also obtained from the Spaniards. . . . each warrior keep[s] one or more horses tyed by a cord to a stake near his lodge both day and night and are always prepared for action at a moments warning. they fight on horseback altogether. . . .

LEWIS / *Thursday August 15th 1805.*

. . . at half after 12 we set out, several of the old women were crying and imploring the great sperit to protect their warriors as if they were going to inevitable distruction. we had not proceeded far before our party was augmented by . . . all the men of the village and a number of women with us. this may serve in some measure to ilustrate the capricious disposition of those people, who never act but from the impulse of the moment. they were now very cheerfull and gay, and two hours ago they looked as sirly as so many imps of satturn. . . .

LEWIS / *Friday August 16th 1805.*

. . . to give them further confidence I put my cocked hat with feather on the chief and my over shirt being of the Indian form my hair deshivled and skin well browned with the sun I wanted no further addition to make me a complete Indian in appearance the men followed my example and we were so[o]n completely metamorphosed. . . .

LEWIS / *Saturday August 17th 1805.—*

. . . Capt Clark arrived with the Interpreter Charbono, and the Indian woman, who proved to be a sister of the Chief Cameahwait. the meeting of those people was really affecting, particularly between Sah-cah-gar-we-ah and an Indian woman, who had been taken prisoner at the same time with her . . . . we next enquired who were chiefs among them. . . . every article about us appeared to excite astonishment in their minds; the appearance of the men, their arms, the canoes, our manner of working them, the black man york and the sagacity of my dog were equally objects of admiration. I also shot my air-gun which was so perfectly incomprehensible that they immediately denominated it the great medicine. . . . To keep indians in a good humour you must not fatiegue them with too much business at one time. . . .

ORDWAY / *Sunday 18th August 1805.*

a clear morning. . . . Capt Lewis bought three horses of the natives. gave a uniform coat and a knife for one and red leggins & a hankerchief & knife for another. a fiew arrow points . . . given for the other . . . .

LEWIS / *Sunday August 18th 1805.*

. . . This day I completed my thirty first year, and conceived that I had in all human probability now existed about half the period which I am to remain in this Sublunary world. I reflected that I had as yet done but little, very little, indeed, to further the hapiness of the human race, or to advance the information of the succeeding generation. I viewed with regret the many hours I have spent in indolence, and now soarly feel the want of that information which those hours would have given me had they been judiciously expended. but since they are past and cannot be recalled, I dash from me the gloomy thought, and resolved in future, to redouble my exertions and at least indeavour to promote those two primary objects of human existence, by giving them the aid of that portion of talents which nature and fortune have bestoed on me; or in future, to live for mankind, as I have heretofore lived for myself.

LEWIS / *Monday August 19th 1805*

. . . notwithstanding their extreem poverty they are not only cheerfull but even gay, fond of gaudy dress and amusements; like most other Indians they are great egotists and frequently boast of heroic acts which they never performed. they are also fond of games of wrisk. they are frank, communicative, fair in dealing, generous with the little they possess, extreemly honest, and by no means beggarly. each individual is his own sovereign master, and acts from the dictates of his own mind; the authority of the Cheif being nothing more than mere admonition supported by the influence which the propiety of his own examplary conduct may have acquired him in the minds of the individuals who compose the band. . . . Sah-car-gar-we-ah had been . . . disposed of before she was taken by the Minnetares, or had arrived to the years of puberty. the husband was yet living with this band. he was more than double her age and had two other wives. he claimed her as his wife but said that as she had had a child by another man, who was Charbono, that he did not want her. . . .

August 20, 1805

CLARK / *August 20th Tuesday 1805*
"So-So-ne" the Snake Indians

Set out at half past 6 oClock and proceeded on (met maney parties of Indians) thro' a hilley Countrey to the Camp of the Indians on a branch [Lemhi River] of the Columbia River, before we entered this Camp a Serimonious hault was requested by the Chief and I smoked with all that Came around, for Several pipes. . . .

ORDWAY / *Tuesday 20th August*

. . . a light frost. . . . the 2 Indians at our Camp behave verry well and their Squaws mend our mockisons . . . and are as friendly as any Savages we have yet seen. our hunters returned in the afternoon but had killed nothing. the game Scearse. . . .

LEWIS / *Wednesday August 21st 1805.*

This morning was very cold. . . . the ink freizes in my pen. . . . neither of the hunters returned this evening and I was obliged to issue pork and corn. . . . some of the dressy young men orniment the tops of their mockersons with the skins of polecats and trale the tail of that animal on the ground at their heels as they walk. . . . They seldom wear the beads . . . about their necks . . . . the men and women were them suspended from the ear in little bunches or intermixed with triangular peices of the shells of the perl oister. the men also were them in . . . the hare . . . of the crown . . . to which they sometimes make the addition of the wings and tails of birds. . . . friends of theirs they say . . . possess a much greater number of horses and mules than they do . . . ; or using their own figure that their horses and mules are as numerous as the grass of the plains. . . .

CLARK / *August 21st Wednesday 1805*

. . . The women are held more sacred among them than any nation we have seen and appear to have an equal Shere in all conversation, which is not the Case in any other nation I have seen. their boys and girls are also admitted to speak except in Councels, the women doe all the drugery . . . . The men who passed by the forks informed me that the S W. fork was double the Size of the one I came down, and I observed that it was a handsom river . . . I shall in justice to Capt Lewis who was the first white man ever on this fork of the Columbia Call this Louis's river. . . .

August 21, 1805

August 23, 1805

August 23, 1805

August 23, 1805

ORDWAY / *Thursday 22nd August 1805.*

... our Intrepter his wife and one tribe of the Snake nation of Indians arived here on horse back . . . . they have come to trade horses with us. Capt Lewis counciled with them made two of their principal men chiefs & gave the meddles, and told them in council that the chief of the 17 great nations of America had sent us to open the road and know their wants, &.C. and told them that . . . we wanted in return their beaver and other Skins if they would take care to save them, &.C. Capt Lewis traded with them and bought three fine horses and 2 half breed mules for a little Marchandize . . . . they appear verry kind and friendly   do not offer to steel or pilfer any thing from us. we trade any usless article . . . . we lend them any thing they want and they are verry careful to return the Same. . . .

CLARK / *August 23rd Friday 1805*

... I Deturmined to delay the party here and with my guide and three men proceed on down to examine if the [Salmon] river continued bad or was practiable . . . . the Mountains Close and is a perpendicular Clift on each Side, and Continues for a great distance and ... the water runs ... foaming & roreing thro rocks in every direction, So as to render the passage of any thing impossible. . . . the rocks & rapids below, at no great distance & The Hills or mountains were . . . like the Side of a tree Streight up.

August 23, 1805

Lewis / Friday August 23rd 1805.

... the Indians pursued a mule buck near our camp I saw this chase for about 4 miles   it was really entertaining, there were about twelve of them in pursuit of it on horseback, they finally rode it down and killed it. ... we found no axes nor hatchets among them; what wood they cut was done either with stone or Elk's horn. the latter they use always to rive or split their wood. their culinary eutensils exclusive of the brass kettle before mentioned consist of pots in the form of a jar made either of earth, or of a white soft stone which becomes black and very hard by birning, and is found in the hills near the three forks of the Missouri between Madison's and Gallitin's rivers. they have also spoons made of the Buffaloe's horn and those of the Bighorn. Their bows are made of ceader or pine and have nothing remarkable about them. the back of the bow is covered with sinues and glue and is about 2½ feet long. much the shape of those used by the Siouxs Mandans Minnetares &c. . . .

Capt. Clark set out this morning very early and poroceeded but slowly ... along the steep side of a mountain over large irregular and broken masses of rocks . . . . Capt Clark . . . now perfictly satisfyed as to the impracticality of this rout either by land or water . . . .

Ordway / Saturday 24th August 1805.

... we had ab^t 20 horses loaded with baggage and Set out about 12 oClock on our journey to cross the the dividing mountains. . . .

Lewis / Saturday August 24th 1805.

... these Indians soon told me that they had no more horses for sale and I directed the party to prepare to set out. I had now nine horses and a mule, and two which I had hired made twelve these I had loaded and the Indian women took the ballance of the baggage. . . . it will require at least 25 horses to convey our baggage along such roads as I expect we shall be obliged to pass in the mountains. I had now the inexpressible satisfaction to find myself once more under way with all my baggage and party. an Indian had the politeness to offer me one of his horses to ride which I accepted with cheerfullness as it enabled me to attend better to the march of the party. . . .

Clark / August 24th Satturday 1805

... I wrote a letter to Capt Lewis informing him of the prospects before us and information recved of my guide which I thought favourable &c. & Stating two plans one of which for us to pursue &c. and despatched one man & horse and directed the party to get ready to march back, every man appeared disheartened from the prospects of the river, and nothing to eate . . . .

The plan I stated to Capt Lewis if he agrees with me we shall adopt is. to procure as many horses (one for each man) if possible and to hire my present guide who I sent on to him to interigate thro' the Intpt^r and proceed on by land to Some navagable part of the Columbia River, or to the Ocean, depending on what provisions we can procure by the gun aded to the Small Stock we have on hand depending on our horses as the last resort.

August 23, 1805

LEWIS / *Monday August 26th 1805.*
... we collected our horses and set out at sunrise. we soon arrived at the extreem source of the Missouri; here I halted a few minutes, the men drank of the water and consoled themselves with the idea of having at length arrived at this long wished for point. ...

ORDWAY / *Tuesday 27th August 1805.*
... in the evening the natives had a war dance .... they were verry merry but did not dance so regular as the Indians on the Missourie. their women sang with them, but did not dance any   they tell us that Some of their horses will dance but they have not brought them out yet. ...

CLARK / *August 29th Thursday 1805—*
... I Spoke to the Indians on various Subjects endeavoring to impress on theire minds the advantage it would be to them for to sell us horses and expedite the [our] journey the nearest and best way possibly ....

I purchased a horse for which I gave my Pistol 100 Balls Powder & a Knife. our hunters Killed 2 Deer near their Camp to day 2 yesterday & 3 the day before, this meet was a great treat to me as I had eate none for 8 days past.

August 29, 1805

August 31, 1805

August 31, 1805

ORDWAY / *Friday 30th August 1805.*

a fine morning. we got up all our horses. bought 8 more. have now got 30 in all. we got our loads ready. the guide who has engaged with us to go on to the ocean tells us . . . the road to the North of the River is rough and mountaineous but s[d] he could take us . . . to where the tide came up and Salt water. . . . went about 10 miles and Camped . . . .

CLARK / *August 31st 1805 Satturday.*

. . . I met an Indian on horse back who fled with great Speed to Some lodges below & informed them that the Enemies were Coming down, arm[d] with guns &c. the inhabitents of the Lodges indisceved him, we proceeded on the road on which I had decended as far as the 1st run [Tower Creek] below & left the road & Proceeded up the Run in a tolerable road 4 miles & Encamped in Some old lodges . . . . Proceeded on 22 miles to Day, 4 miles of which up a run

    Course and Distance by land from the Columbia River 14 miles below the forks.
                    August 31st 1805.

N. 35° E 2 miles up Tower Creek to a hill
N. 10° E 2
      $\overline{4}$ d°   d°   d°   passed remarkable rock resembling Pirimids on the Left Side

September 3, 1805

CLARK / *September 3rd Tuesday 1805—*

... Sent 2 men back with the horse on which Capt Lewis rode for the load left back last night which detained us untill 8 oClock at which time we Set out. The Country is timbered with Pine Generally, the bottoms have a variety of Srubs & the fur trees in Great abundance, hills high & rockey on each Side, in the after part of the day the high mountains closed the Creek on each Side and obliged us to take on the steep Sides of those Mountains ... Encamped on a branch of the Creek we assended after crossing Several Steep points & one mountain ....

WHITEHOUSE / *Tuesday 3rd Sep^t 1805.*

... the mountains was So Steep and rockey that Several of the horses fell back among the rocks and was near killing them. Some places we had to cut the road through thickets of bolsom fer .... we Camped after a dissagreeable days march of only 11 miles with much fatigue and hunger as nothing has been killed this day only 2 or 3 fessents, and have no meat of any kind. Set in to raining hard at dark So we lay down and Slept, wet hungry and cold. Saw Snow on the tops of Some of these mountains this day.

CLARK / *September 4th Wednesday 1805—*

... we met a party of the Tushepau [Flathead] nation, of 33 Lodges   about 80 men 400 Total and at least 500 horses ... those Indians are well dressed with Skin shirts & robes ... & light complected more So than Common for Indians, The chief harangued untill late at night, Smoked in our pipe and appeared Satisfied. I was the first white man who ever wer on the waters of this river [Fish Creek].

WHITEHOUSE / *Wednesday 4th Sep^t 1805.*

... they have between 4 and 500 well looking horses now feeding in this valley or plain in our view. they received us as friends and appeared to be glad to See us.... they tell us that we can go in 6 days to where white traders come and that they had Seen bearded men who came [from] a river to the North of us 6 days march ....

ORDWAY / *Thursday 5th Sep^t 1805.*

... our officers bought 12 horses from them [Flatheads] .... found it verry troublesome Speaking to them as all they Say to them has to go through Six languages, and hard to make them understand....

they appear to us as though they had an Impediment in their Speech or brogue on their tongue....

GASS / Thursday [September] 5th. [1805]
This was a fine morning with a great white frost. The Indian dogs are so hungry and ravenous, that they eat 4 or 5 pair of our mockasons last night. We remained here all day, and recruited our horses to 40 and 3 colts; and made 4 or 5 of this nation of Indian chiefs. They are a very friendly people ... but they have nothing to eat, but berries, roots, and such articles of food.... They are the whitest Indians I ever saw.

WHITEHOUSE / Friday 6th Sept 1805.
... we take these Savages to be the Welch Indians if their be any Such from the Language. So Capt Lewis took down the names of everry thing in their Language, in order that it may be found out whether they are or whether they Sprang or origenated first from the welch or not....

CLARK / September 7th Satturday 1805—
A Cloudy & rainie Day the greater Part of the day dark & Drisley   we proceeded on down the river thro a Vallie ... from 1 to 2 miles wide ....

September 4, 1805

September 7, 1805

September 9, 1805

September 9, 1805

LEWIS / *Monday September 9th 1805.*

Set out at 7 A.M. this morning and proceeded down the Flathead [Bitter Root] river leaving it on our left, the country in the valley of this river is generally a prarie and from five to 6 miles wide the growth is almost altogether pine principally of the long-leafed kind, with some spruce and a kind of furr resembleing the scotch furr. near the wartercourses we find a small proportion of the narrow leafed cottonwood; some redwood honeysuckle and rosebushes form the scant proportion of underbrush to be seen. . . . we continued our rout down the W. side of the river about five miles . . . and encamped on a large creek which falls in on the West. as our guide inform me that we should leave the river at this place and the weather appearing settled and fair I determined to halt the next day rest our horses and take some scelestial Observations. we call this Creek Travellers Rest. it is about 20 yards wide a fine bould clear runing stream. . . . we estimate our journey of this day at 19. M.

LEWIS / *Tuesday September 10th 1805*

The Minetares informed us that there wass a large river west of, and at no great distance from the sources of Medicine [Sun] river, which passed along the Rocky Mountains from S. to N.   this evening one of our hunters [Pvt. John Colter] returned accompanyed by three men of the Flathead nation whom he had met in his excurtion up travellers rest Creek. on first meeting him the Indians were alarmed and prepared for battle with their bows and arrows, but he soon relieved their fears by laying down his gun and advancing towards them. the Indians were mounted on very fine horses of which the Flatheads have a great abundance; that is, each man in the nation possesses from 20 to a hundred head. our guide could not speak the language of these people but soon engaged them in conversation by signs or jesticulation, the common language of all the Aborigines of North America, it is one understood by all of them and appears to be sufficiently copious to convey with a degree of certainty the outlines of what they wish to communicate. in this manner we learnt from these people that two men which they supposed to be of the Snake nation had stolen 23 horses from them and that they were in pursuit of the theaves. they told us they were in great hast, we gave them some boiled venison, of which the [y] eat sparingly. the sun was now set, two of them departed after receiving a few small articles which we gave them, and the third remained, having agreed to continue with us as a guide, and to introduce us to his relations whom he informed us were numerous and resided in the plain below the mountains on the columbia river, from whence he said the water was good and capable of being navigated to the sea; that some of his relation[s] were at the sea last fall and saw an old whiteman who resided there by himself and who had given them some handkerchiefs such as he saw in our possession. he said it would require five sleeps

September 10, 1805

September 10, 1805

CLARK / *September 11th Wednesday 1805—*

... Sent out the hunters to hunt in advance as usial. ... we proceeded on up the Creek [Travelers rest] *on the right* [hand] *Side* [left bank] *thro a narrow valie and good road for 7 miles ... the mountains on the left high & Covered with Snow. The day Verry worm*

WHITEHOUSE / *Wednesday 11th Sept 1805.*

a beautiful pleasant morning. ... passed a tree on which was a nomber of Shapes drawn on it with paint by the natives. a white bear Skin hung on the Same tree. we Suppose this to be a place of worship among them. Came about 7 miles ....

September 11, 1805

September 12, 1805

CLARK / *September 12th Thursday 1805.*

... passed a Fork on the right on which I saw near an old Indian encampment a Swet (Sweat) house Covered with earth ....

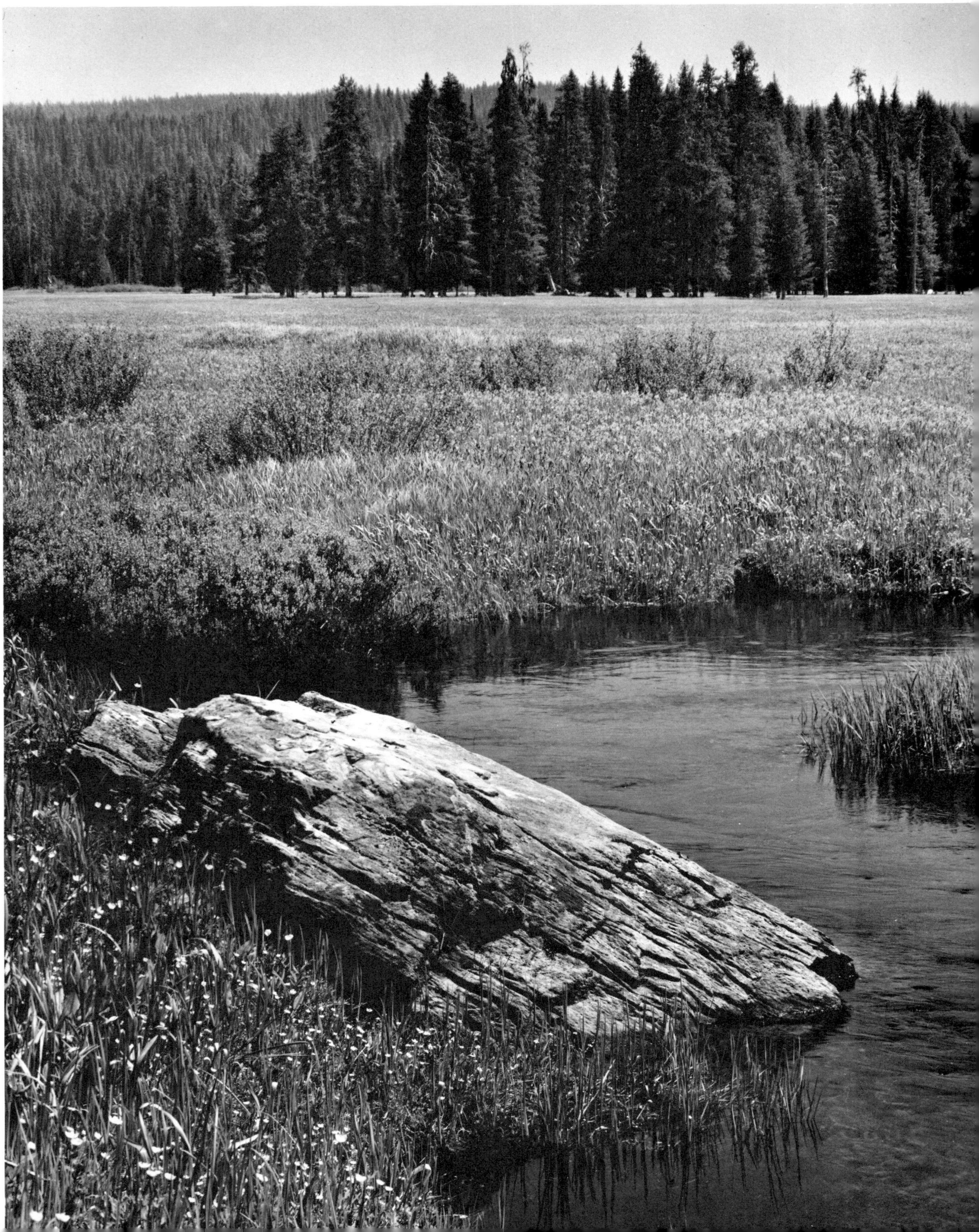

September 14, 1805

CLARK / *September 13th Wednesday (Friday) 1805—*

... *The pine Countrey falling timber &c. &c. Continue. . . . and we proceeded over a mountain to the head of the Creek which we left to our left and at 6 miles from the place I nooned it, we fell on a Small Creek from the left which Passed through open glades* [Packer Meadow] . . . .

September 13, 1805

CLARK / *September 14th Thursday (Saturday) 1805*

... *Crossed a high mountain on the right of the Creek for 6 miles to the forks of the Glade Creek (one of the heads of the Koos koos kee) . . . . Encamped opposit a Small Island at the mouth of a branch on the right side of the river which is at this place 80 yards wide, Swift & Stoney, here we were compelled to kill a Colt for our men and Selves to eat . . . The Mountains which we passed to day much worst than yesterday the last excessively bad & thickly Strowed with falling timber & Pine Spruce fur Hackmatak & Tamerack, Steep & Stoney   our men and horses much fatigued*

September 15, 1805

CLARK / *Wednesday (Sunday) Sepr 15th 1805*

... here the road leaves the river ... and assends a mountain winding in every direction to get up the Steep assents ... 4² miles up the mountain I found a Spring and halted for the rear to come up and to let our horses rest & feed, [in] about 2 hours the rear of the party came up much fatigued & horses more so, Several horses Sliped and ... the one which Carried my desk & Small trunk Turned over & roled down a mountain for 40 yards & lodged against a tree, broke the Desk the horse escaped and appeared but little hurt ... we proceeded on up the mountain Steep & ruged as usial, more timber near the top, when we arrived at the top As we Conceved, we could find no water and Concluded to Camp and make use of the Snow we found on the top to cook the remn^s of our Colt & make our Supe, evening verry cold and cloudy....

From this mountain I could observe high ruged mountains in every direction as far as I could see. with the greatest exertion we could only make 12 miles up this mountain ....

September 15, 1805

September 15, 1805

September 15, 1805

September 16, 1805

WHITEHOUSE / *Monday 16th Sept 1805.*

when we awoke this morning to our great Surprise we were covred with Snow . . . . we mended up our mockasons. Some of the men without Socks raped rags on their feet, and loaded up our horses and Set out without anything to eat . . . . kept on the ridge of the mountain . . . .

CLARK / *Saturday (Monday) Sept: 16th 1805*

. . . I have been wet and as cold in every part as I ever was in my life, indeed I was at one time fearfull my feet would freeze in the thin Mockirsons which I wore, after a Short Delay in the middle of the Day, I took one man and proceeded on as fast as I could about 6 miles . . . and built fires for the party . . . we Encamped at this Branch in a thickly timbered bottom which was scurcely large enough for us to lie leavil, men all wet cold and hungary. Killed a Second Colt which we all Suped hartily on and thought it fine meat. . . .

CLARK / *Sunday (Tuesday) 17th Septr 1805—*

. . . no Snow in the Vallies Killed a fiew Pheasents which was not sufficient for our Supper which compelled us to kill Something, a Coalt being the most useless part of our Stock he fell a Prey to our appetites. The after part of the day fare, we made only 10 miles to day two horses fell & hurt themselves very much   we Encamped on the top of a high Knob of the mountain at a run passing to the left. . . .

CLARK / *Monday (Wednesday) 18th Septr 1805—*

a fair morning cold   I proceeded on in advance with Six hunters (and let it be understood that my object was) to try and find deer or Something to kill (& send back to the party) (The want of provisions together with the dificul[t]y of passing those emence mountains dampened the sperits of the party which induced us to resort to Some plan of reviving ther sperits. I deturmined to take a party of the hunters and proceed on in advance to Some leavel Country, where there was game kill Some meat & send it back &c.)

September 17, 1805

September 18, 1805

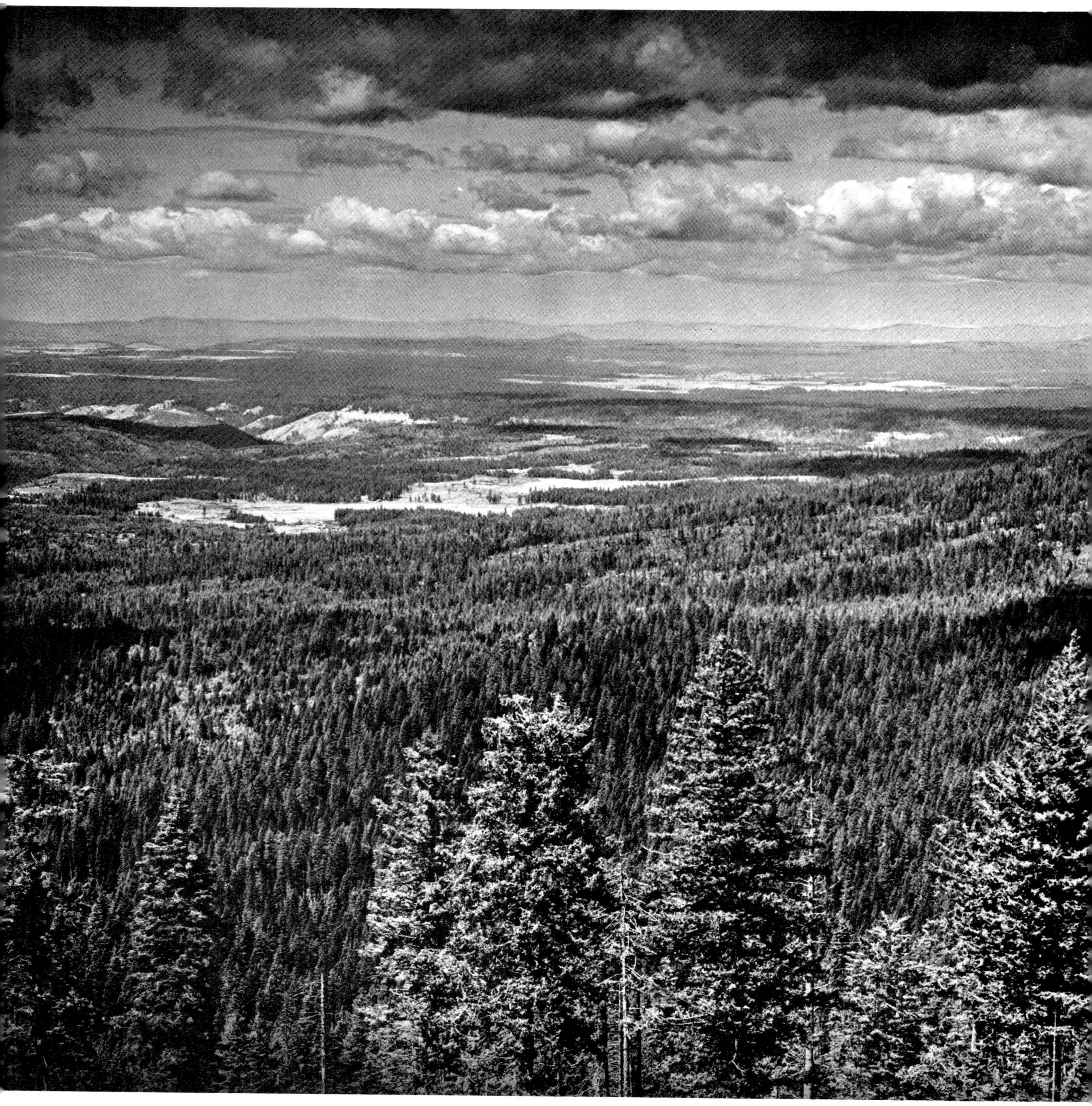

LEWIS / *Thursday September 19th 1805.*

Set out this morning a little after sun rise ... and we to our inexpressable joy discovered a large tract of Prairie country lying to the S. W. and widening as it appeared to extend to the W.   through that plain the Indian informed us that the Columbia river, (in which we were in surch) run. this plain appeared to be about 60 Miles distant, but our guide assured us that we should reach it's borders tomorrow   the appearance of this country, our only hope for subsistance greately revived the sperits of the party already reduced and much weakened for the want of food. ...

CLARK / *Wednesday (Friday) 20th September 1805*

I set out early and proceeded on through a Countrey as ruged as usial   passed over a low mountain into the forks of a large Creek which I kept down 2 miles and assended a high Steep mountain leaveing the Creek to our left hand   passed the head of several dreans on a divideing ridge, and at 12 miles decended the mountain to a leavel pine Countrey proceeded on through a butifull Countrey for three miles to a Small Plain in which I found maney Indian lodges ....

September 19, 1805

CLARK / *Wednesday (Friday) 20th September 1805*

...(Soon after) a man Came out to meet me, [with great caution] & Conducted me [us] to a large Spacious Lodge which he told me (by Signs) was the Lodge of his great Chief who had Set out 3 days previous with all the Warriers of the nation to war on a South West derection & would return in 15 or 18 days. the fiew men that were left in the Village and great numbers of women geathered around me with much apparent signs of fear, and ap^r pleased they those people gave us a Small piece of Buffalow meat, Some dried Salmon beries & roots in different States, Some round and much like an onion which they call Pas she co [quamash. the Bread or Cake is called Pas-shi-co] Sweet, of this they make bread & Supe they also gave us, the bread made of this root all of which we eate hartily .... They call themselves Cho pun-nish or Pierced noses Their diolect appears verry different from the flat heads, [Tushapaws], altho origineally the Same people, They are darker than the Flat heads I have seen [Tushapaws Their] dress Similar, with more beads white and blue principally, brass & Copper in different forms, Shells and ware their haire in the Same way. they are large Portley men Small women & handsom featured ....

I find myself verry unwell all the evening from eateing the fish & roots too freely ....

CLARK / *Thursday (Saturday) 21 Septr 1805*

... we did not arrive at the Camp of the Twisted hare but opposit, untill half past 11 oClock P.M. found at this Camp five Squars & 3 children.... I found him a Cheerfull man with apparant siencerity. I gave him a Medal &c. and Smoked untill 1 oClock a.m. and went to Sleep. The Country from the mountains to the river hills is a leavel rich butifull Pine Countrey badly watered, thinly timbered & covered with grass. The weather verry worm after decending into the low Countrey, the river hills are Verry high & Steep, Small bottoms to this little river which is Flat head [Clearwater] & is 160 yards wide and Sholey. This river is the one we killed the first Coalt on near a fishing were.

I am verry sick to day and puke which relive me

September 20, 1805

September 20, 1805

September 21, 1805

CLARK / Friday (Sunday) 22nd Septr 1805

... walked upto the 2d Village where I found Capt Lewis & the party Encamped, much fatigued, & hungery, much rejoiced to find something to eate ....
   The planes appeared covered with Spectators viewing the white men and the articles which we had .... I precured maps of the Country & river with the Situation of Indians, Towns ....

ORDWAY / Monday 23rd Sept 1805.

   a fair morning. we purchased considerable of Sammon and commass roots from the natives [Nez Percés] .... also they are fond of any kind of marchandize, but the blue beeds they want mostly. our officers gave the chiefs of this nation a flag a meddle .... these natives have a great many horses and live well. ...

CLARK / Sunday (Tuesday) 24th Sepr 1805

... Capt Lewis scercely able to ride on a jentle horse which was furnished by the Chief,   Several men So unwell that they were Compelled to lie on the Side of the road for Some time   others obliged to be put on horses. I gave rushes Pills to the Sick this evening. Several Indians follow us.

September 26, 1805

CLARK / *Tuesday (Thursday) 26th Septr 1805*

Set out early and proceeded on down the river to a bottom opposit the forks of the river on the South Side and formed a Camp.... I had the axes distributed and handled and men apotn^cd [apportioned] ready to commence building canoes on tomorrow, our axes are small & badly calculated to build Canoes of the large Pine, Capt Lewis Still verry unwell, Several men taken Sick on the way down, I administered Salts Pils Galip, [jalap] Tarter emetic &c. I feel unwell this evening ....

WHITEHOUSE / *Thursday 26th Sept 1805.*

... we formed our Camp in a narrow plain on the bank of the River. made a pen of pine bushes around the officers lodge, to put all our baggage in. Some of the natives followed us with droves of horses.... men Sick with the relax ....

CLARK / *(Friday) Sunday 29th Septr 1805*

... men Sick as usial, all the men (that are) able to (at) work, at the Canoes   Drewyer killed 2 Deer Colter killed 1 Deer ....

CLARK / *October 1st Monday (Tuesday) 1805—*

... laid out a Small assortment of such articles as those Indians were fond of to trade with them for Some provisions (they are remarkably fond of Beeds) nothin to eate except a little dried fish which they men complain of as working of them as (as much as) a dost of Salts. Capt Lewis getting much better. Several Indians visit us from the different tribes below. Some from the main South fork. our hunters killed nothing to day worm evening

CLARK / *October 5th Friday Saty 1805*

... had all our horses 38 in number Collected and branded   Cut off their fore top and delivered them to the 2 brothers and one son of the Chiefs who intends to accompany us down the river .... they promised to be attentive to our horses untill we Should return.
... Capt Lewis & myself eate a Supper of roots boiled, which Swelled us in Such a manner that we were Scercely able to breath for Several hours. finished and lanced (launched) 2 of our canoes this evening ....

October 5, 1805

CLARK / *October 6th Saturday [Sunday] 1805*

... The river below this forks is Called Kos-kos-kee [Clearwater]   it is Clear rapid with Shoals or Swift places

The open Countrey Commences a fiew miles below this on each side of the river . . . .

CLARK / *October 7th Monday 1805—*

I continue verry unwell but obliged to attend every thing   all the Canoes put into the water and loaded . . . and Set out . . . .
. . . the Indians say . . . the great falls 10 day below, where the white people live &c. . . .

GASS / *[October 7, 1805]*

. . . We had four large canoes; and one small one, to look a-head. About three o'clock in the afternoon we began our voyage down the river . . . .

WHITEHOUSE / *Wednesday 9th Oct. 1805.*

. . . we were obledged to delay and prepare or repair the canoe which got Stove last evening, put the loading marchandize &c out to dry. . . . the natives hang about us as though they wished to Steal or pilfer Something from us So we had to keep 2 Sentinels . . . . after dark we played the fiddle and danced a little. the natives were pleased to see us. one of their women was taken with the crazey fit by our fire. . . .

CLARK / *October 10th Wednesday (Thursday) [1805]*

. . . we landed near 8 Lodges of Indians (Choponnesh) [Nez Percé] . . . . we purchased fish & dogs of those people . . . .
. . . all the Party have greatly the advantage of me . . . they all relish the flesh of the dogs . . . .
The Cho-pun-nish or Pierced nose Indians are Stout likely men, handsom women, and verry dressey in their way,   the dress of the men are a White Buffalow robe or Elk Skin dressed with Beeds which are generally white, Sea Shells & the Mother of Pirl hung to ther hair & on a piece of otter skin about their necks   hair Ceewed in two parsels hanging forward over their Sholders, feathers, and different Coloured Paints which they find in their Countrey Generally white, Green & light Blue. Some fiew were a Shirt of Dressed Skins and long legins & Mockersons Painted . . . .

October 6, 1805

October 11, 1805

The women dress in a Shirt of Ibex or Goat [Argalia] Skins which reach quite down to their anckles with a girdle, their heads are not ornemented, their Shirts are ornemented with quilled Brass, Small peces of Brass Cut into different forms, Beeds, Shells & curious bones &c. The men expose those parts which are generally kept from few [view] by other nations but the women are more perticular than any other nation which I have passed [in secreting the parts] . . .

CLARK / *October 11th Friday 1805*

. . . at 6 miles [on Snake River] we came too at Some Indian lodges and . . . at this place I saw a curious Swet house . . . with a Small whole to pass in or throw in the hot Stones, which those in[side] threw on as much water as to create a temporature of heat they wished . . . .

CLARK / *October 12th Saturday 1805*

. . . today Country as yesterday open plains, no timber of any kind, a fiew Hackberry bushes & willows excepted . . . So that fire wood is verry Scerce   The hills or assents from the water is faced with a dark ruged Stone. . . .

October 12, 1805

October 13, 1805

CLARK / *October 13th Sunday 1805*

[second draft] ... *passed the Mo: ... below a long bad rapid* [Drewyers (Palouse) River] *in which the water is confined in a Chanel of about 20 yards between rugid rocks for the distance of a mile and a half, and a rapid rockey chanel for 2 miles above. This must be a verry bad place in high water, here is great fishing place ....*

[first draft] *... The wife of Shabono our interpreter we find reconsiles all the Indians, as to our friendly intentions   a woman with a party of men is a token of peace*

CLARK / *October 14th Monday 1805*

a verry Cold morning wind from the West and Cool untill about 12 oClock when it Shifted to the S.W.   at 2-½ miles passed a remarkable rock verry large and resembling the hill [hull] *of a Ship Situated on a Lar^d point at some distance from the assending Countrey   passed rapids at 6 and 9 miles. ...*

October 14, 1805

ORDWAY / *Monday 14th Oct 1805.*

... came to another bad rapid at the head of an Island. the canoe I had charge of ran fast on a rock .... considerable of the baggage washed overboard, but the most of it was taken up below ... She went off of a sudden & left myself and three more standing on the rock half leg deep in the rapid water. ...

CLARK / *October 15th Tuesday 1805*

... Cap! Lewis walked on the plains and informs that he could plainly See a rainge of mountains which bore S.E. & N.W. ... we landed at a parcel of split timber, the timber of a house of Indians ... here we were obliged for the first time to take the property of the Indians without the consent or approbation of the owner. ... We made only 20 miles today, owing to the detention in passing [Snake River] rapids &c.

CLARK / *October 16th Wednesday 1805*

... Set out and proceeded on Seven miles to the junction of this river [the Snake] and the Columbia which joins from the N.W. ... In every direction from the junction of those rivers the countrey is one continued plain low and rises from the water gradually, except a range of high Countrey which runs from S.W. & N.E. and is on the opposit Side about 2 miles distant from the Collumbia and keeping its derection S.W. untill it joins a S.W. range of mountains.

... after we had our camp fixed and fires made, a Chief came from this camp which was about ¼ of a mile up the Columbia river at the head of about 200 men singing and beeting on their drums Stick and keeping time to the musik, they formed a half circle around us and Sung for Some time, we gave them all Smoke, and Spoke to their Chief as well as we could by signs informing them of our friendly disposition to all nations, and our joy in Seeing those of our Children around us, Gave the principal chief a large Medal, Shirt and Handkf.   a 2nd Chief a Meadel of Small size, and to the Cheif who came down from the upper villages a Small Medal & Handkerchief. ...

October 15, 1805

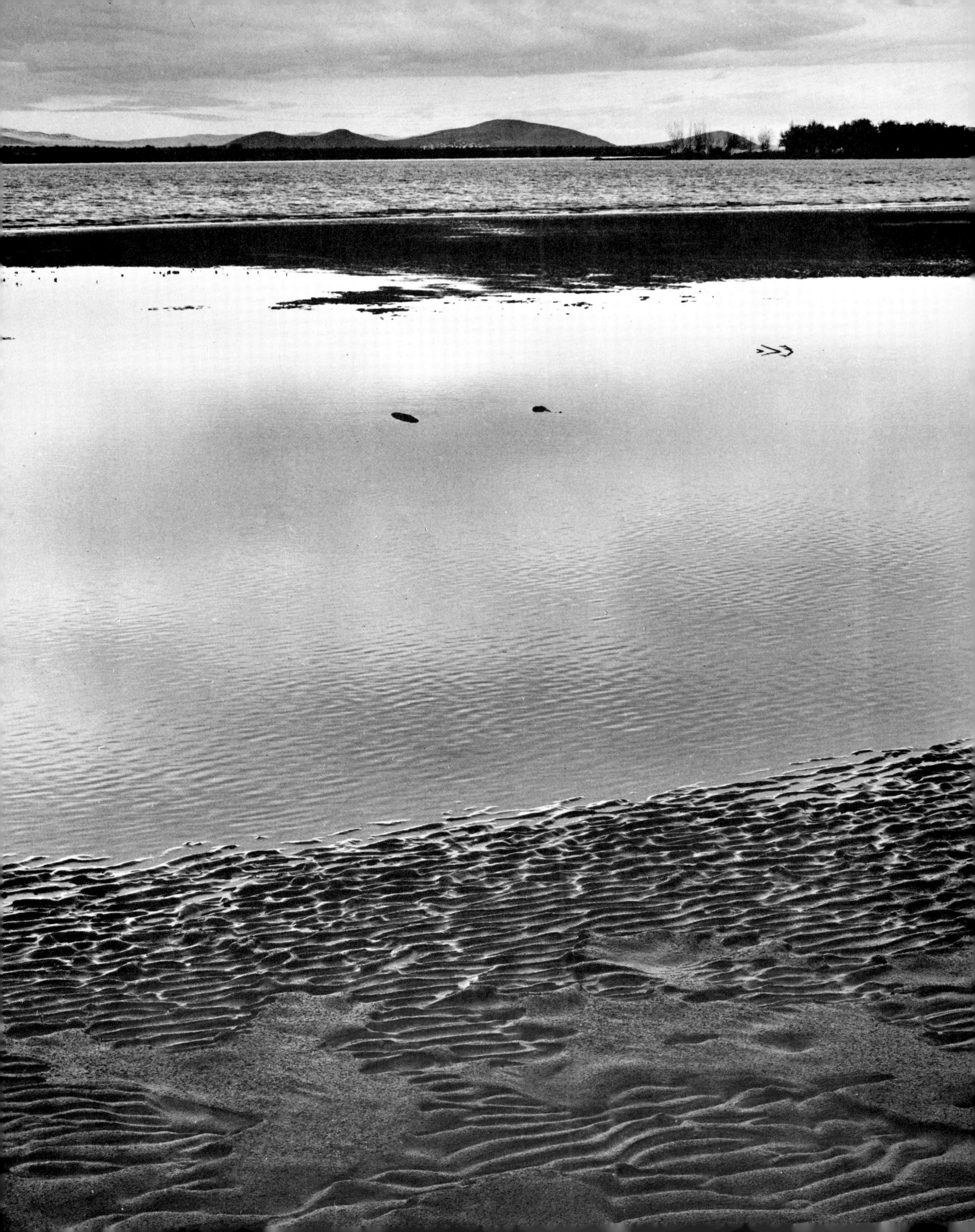

October 18, 1805

ORDWAY / Friday 18th Oct 1805.

a clear pleasant morning. . . . we proceeded on down the great Columbia River which is now verry wide about ¾ of a mile in General   the country in general Smooth plains for about 10 miles down then the barron hills make close to the River on each Side . . . .

October 18, 1805

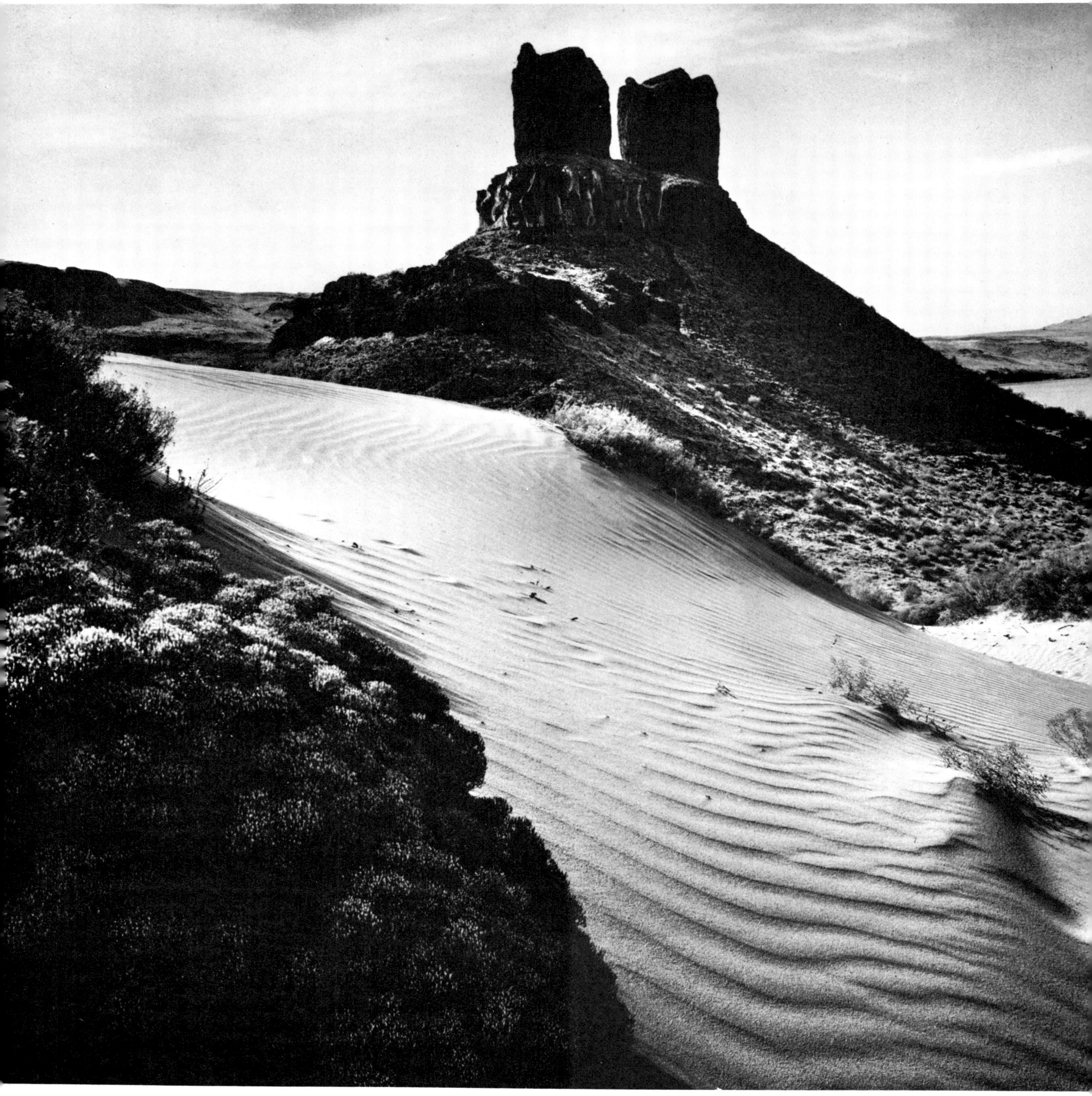

CLARK / October 18th Friday 1805

... Several canoes of Indians came down and joined those with us, we had a council with those in which we informed of our friendly intentions towards them and all other of our red children, of our wish to make a piece between all of our red Children....

The Great Chief [Cuts-sâh-nem] and one of the Chim-nà-pum *nation drew me a sketch of the Columbia above and the tribes of his nation*....

we thought it necessary to lay in a Store of Provisions for our voyage, and the fish being out of Season, we purchased forty dogs for which we gave articles of little value, such as bells, thimbles, knitting pins, brass wire and a few beeds....

Took our leave of the Chiefs and all those about us and proceeded on down the great Columbia river .... the river passes into the range of high Countrey, at which place the rocks project into the river from the high clifts which is on the Lar^d Side about ⅓ of the way across and those of the Star^d Side about the same distance, the countrey is bordered with black rugid rocks .... saw a mountain bearing S. W. conocal form Covered with Snow....

October 18, 1805

October 19, 1805

October 21, 1805

CLARK / *October 19th Saturday* [1805]

[*first draft*] *The Great Chief 2d Chief and a Chief of a band below came and smoked with us   we gave a meadel a string of Wampom & handkerchef to the Great Chief by name* Yel-lep-pit . . . . *the Chief requested us to stay untill 12   we excused our selves and set out at 9 oClock*

Course

SW.   14 *miles to a rock in a Lard resembling a hat just below a rapid* . . . .

ORDWAY / *Monday 21st Oct 1805.*

. . . proceeded on   passed River hills and cliffs of rocks on each side. passed over a number of bad rockey rapids where the River is nearly filled with high dark couloured rocks the water divided in narrow deep channels, bad whorl pools. . . . we came about 32 miles . . . .

CLARK / *October 21ˢᵗ Monday 1805*

... passᵈ a Small Island at 5½ miles a large one 8 miles in the middle of the river, some rapid water at the head and Eight Lodges of nativs . . . we came too at those lodges, bought some wood and brackfast, Those people recived us with great kindness . . . their employments customs, Dress and appearance Similar to those above, Speak the Same language, here we Saw two scarlet and a blue cloth blankets, also a Salors Jacket. . . .

. . . the river is crouded with rocks in every direction, after Passing this dificult rapid to the mouth of a Small river on the Larboard Side 40 yards wide descharges but little water at this time, and appears to take its Sourse in the open plains to the S. E. . . . great numbˢ of Indians came in canoes to View us at this place, after passing this rapid which we accomplished without loss; winding through between the hugh rocks for about 2 miles. . . .

October 21, 1805

CLARK / *October 21st Monday 1805*

... *(from this rapid the Conical mountain is S.W. which the Indians inform me is not far to the left of the great falls; this I call the* Timm *or falls mountain [Mount Hood] it is high and the top is covered with snow) imediately below the last rapids there is four Lodges of Indians....*

... proceeded on about two miles lower and landed and encamped near five Lodges of nativs ... we purchased a little wood to cook our Dog meat and fish; those people did not receive us at first with the same cordiality of those above, they appear to be the Same nation Speak the Same language with a little curruption of maney words Dress and fish in the same way, all of whome have pierced noses *and the men when Dressed ware a long taper'd piece of Shell or beed put through the nose....*

The probable reason of the Indians residing on the Star^d *Side of this as well as the waters of Lewis's River is their fear of the Snake Indians who reside, as they nativs say on a great river to the South, and are at war with those tribes.... One of our party J. Collins presented us with Some verry good beer made of the Pa-shi-co-quar-mash bread, which bread is the remains of what was laid in as a part of our Stores of Provisions, at the first flat heads or Cho-pun-nish Nation....*

ORDWAY / *Tuesday 22nd Oct 1805.*

... this Is^d *is about 4 miles in length and high rough & rockey. a Short distance below we came to the first falls of the Columbia River....*

October 21, 1805

October 22, 1805

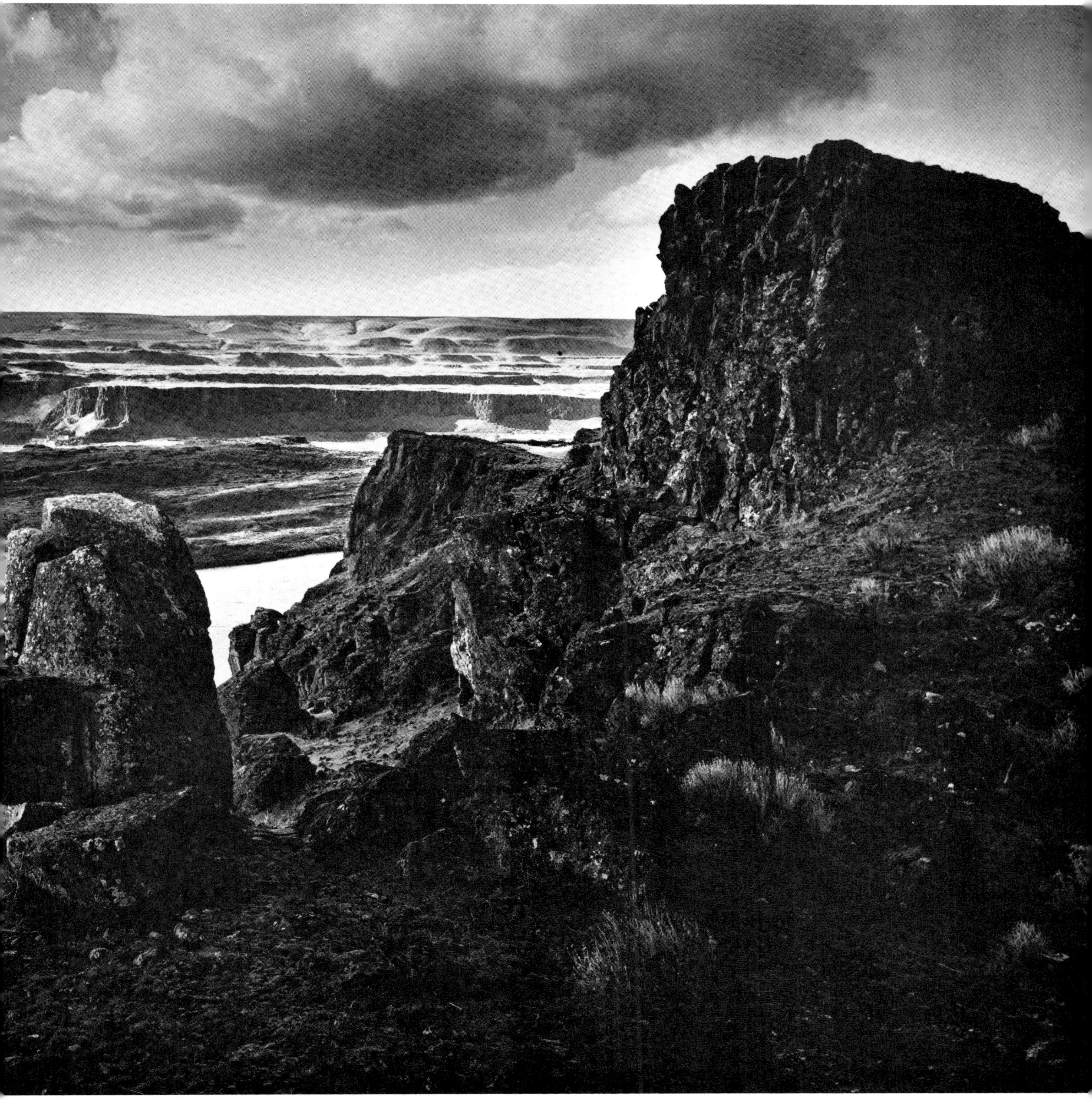

October 22, 1805

CLARK / *October 22ᵈ Tuesday 1805*

... opposite on the Starᵈ Side is 17 Lodges of the nativs   we landed and walked down accompanied by an old man to view the falls, and the best rout for to make a portage which we Soon discovered was much nearest on the Starᵈ Side, and the distance 1200 yards one third of the way on a rock, about 200 yards over a loose Sand collected in a hollar blown by the winds.... we returned droped down to the head of the rapids and took every article except the Canoes across the portag[e] where I had formed a camp on [an] ellegable Situation for the protection of our Stores from thieft, which we were more fearfull of, than their arrows.... at and about their Lodges I observe great numbers of Stacks of pounded Salmon neetly preserved in the following manner, i.e. after [being] suffi[c]ently Dried it is pounded between two Stones fine, and put into a speces of basket neetly made of grass and rushes better than two feet long and one foot Diamiter, which basket is lined with the Skin of Salmon Stretched and dried for the purpose.... Great quantities as they inform us are sold to the whites people who visit the mouth of this river as well as to the nativs below....

GASS / *Wednesday [October] 23rd. [1805]*

... in high water there is nothing but a rapid, and the salmon can pass up without difficulty. The reason of this rise in the water below the falls is, that for three miles down, the river is so confined by rocks (being not more than 70 yards wide) that it cannot discharge the water as fast as it comes over the falls, until what is deficient in breadth is made up in depth. About the great pitch the appearance of the place is terrifying, with vast rocks, and the river below the pitch, foaming through different channels.

CLARK / *October 23ᵈ Wednesday 1805*

... landed Safe with all the canoes at our Camp below the falls by 3 oClock P.M.... every man of the party was obliged to Strip naked dureing the time of taking over the canoes, that they might have an oppertunity of brushing the flees off their legs and bodies.... one of the old Chiefs who had accompanied us from the head of the river, informed us that he herd the Indians Say that the nation below intended to kill us.... as we are at all times & places on our guard, are under no greater apprehention than is common.

we purchased 8 Small fat dogs for the party to eate....

CLARK / *October 24ᵗʰ Thursday 1805*

... At 9 oClock a.m. I Set out with the party and proceeded on down a rapid Stream of about 400 yards wide   at 2-½ miles the river widened into a large bason.... I could See the dificuelties we had to pass for Several miles below.... The whole of the current of this great river must at all Stages pass thro' this narrow chanel of 45 yards wide. as the portage of our canoes ... would be impossible with our Strength, and the only danger in passing thro those narrows was the whorls and swills [swells] arriseing from the Compression of the water, and which I thought (as also our principal watermen Peter Crusat) by good Stearing we could pass down Safe, accordingly I deturmined to pass through this place notwithstanding the horrid appearance of this agitated gut swelling, boiling & whorling in every direction, (which from the top of the rock did not appear as bad as when I was in it; however we passed Safe to the astonishment of all the Indˢ of the last Lodges who viewed us from the top of the rock....

October 24, 1805

October 25, 1805

CLARK / *October 25th Friday 1805*

... we proceeded on down   the water fine, rocks in every derection for a fiew miles when the river widens and becoms a butifull jentle Stream of about half a mile wide,   Great numbers of the Sea orter [or Seals] about those narrows and both below and above....

CLARK / *October 26th Saturday 1805.*

... all our articles we have exposed to the Sun to Dry; and the Canoes drawn out and turned up. maney of our Stores entirely spoiled by being repeetedly wet, ...
   The nations in the vicinity of this place is at War with the Snake Indians....

WHITEHOUSE / *Wednesday 30th Oct. 1805.*

... one half mile above the falls is a village of about 10 well looking cabbins covred with bark.... these Savages were Surprized to See us   they Signed to us that they thought that we had rained down out of the clouds....

CLARK / *October 31st Thursday 1805*

... I could not see any rapids below in the extent of my view which was for a long distance down the river, which from the last rapids widened and had everry appearance of being effected by the tide [this was in fact the first tide water] I deturmind to return to camp 10 miles distant,   a remarkable high detached rock Stands in a bottom on the Star.d Side near the lower point of this Island on the Star.d Side about 800 feet high and 400 paces around, we call the Beaten [Beacon] rock....

October 31, 1805

October 25, 1805

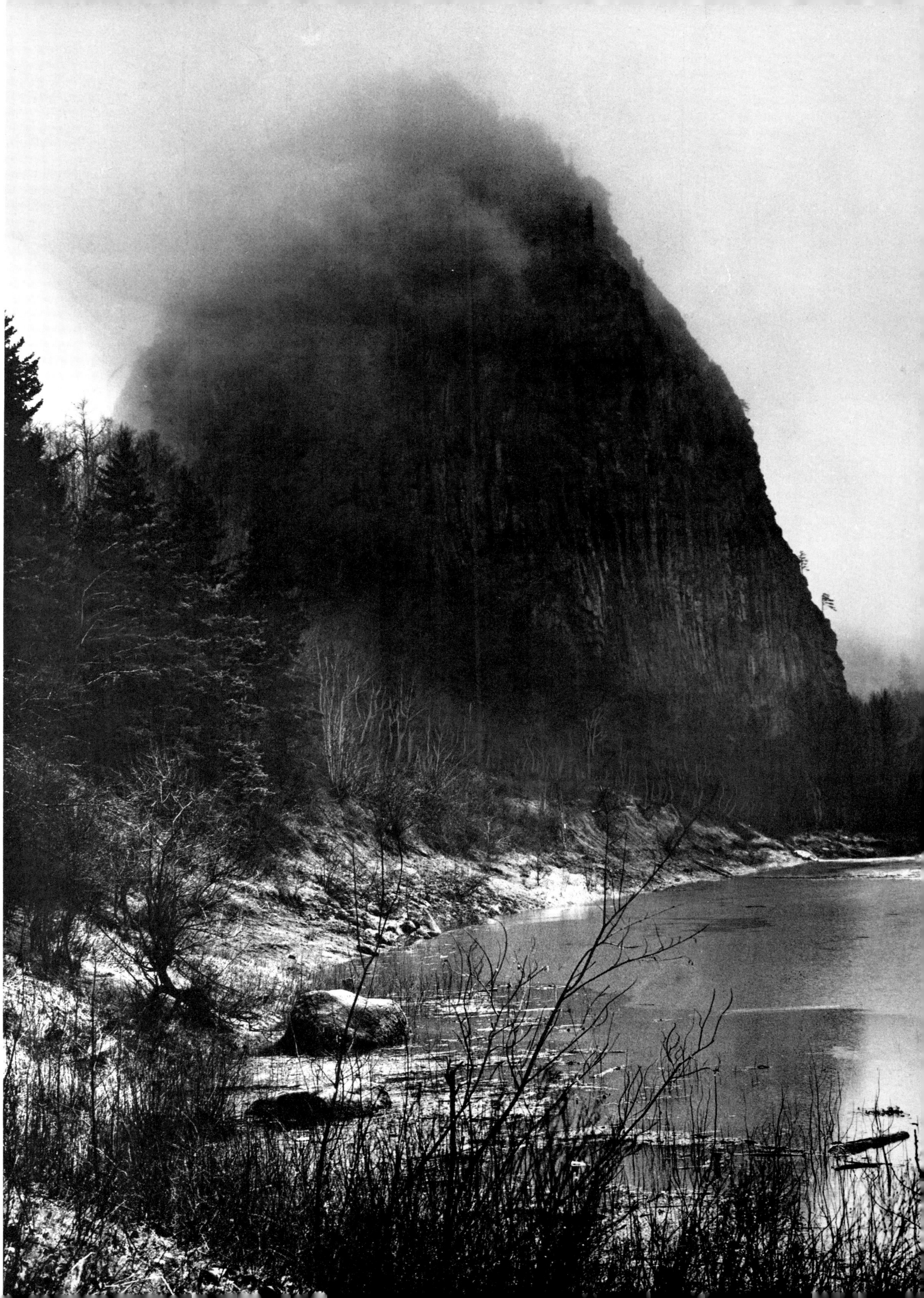

WHITEHOUSE / *Sunday 3rd Nov 1805.*

a foggy morning.... we Saw a high round mountain on the Lar^d Side which we expect is the Same we Saw ab^o the great falls and the Same that Lieu^t Hood gave an account off (it is nearly cov^d with Snow).... towards evening we met Several Indians in a canoe who were going up the River. they Signed to us that in two Sleeps we Should See the Ocean vessels and white people &c. &c. the Country lower and not So mountanous the River more handsome ....

November 3, 1805

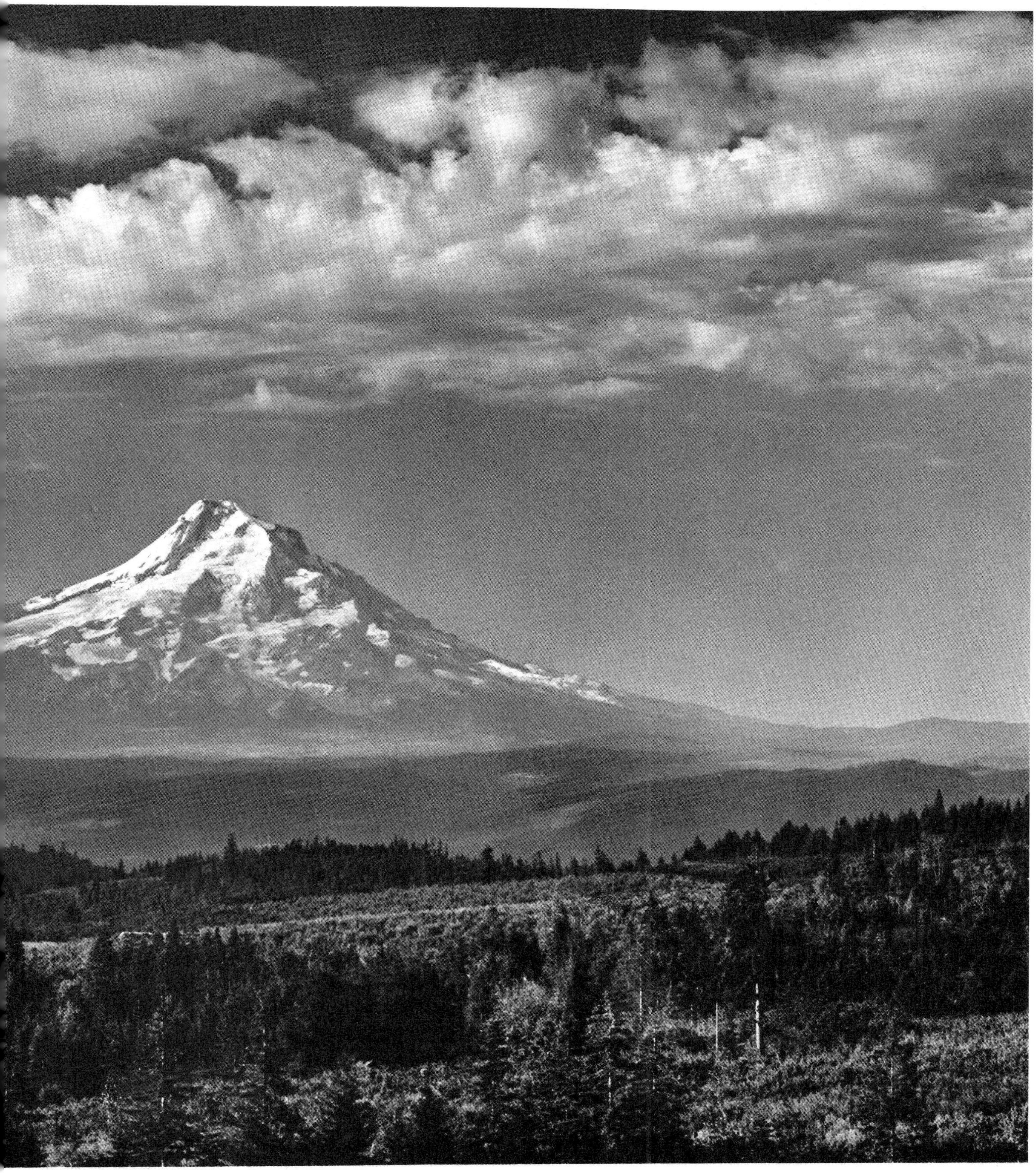

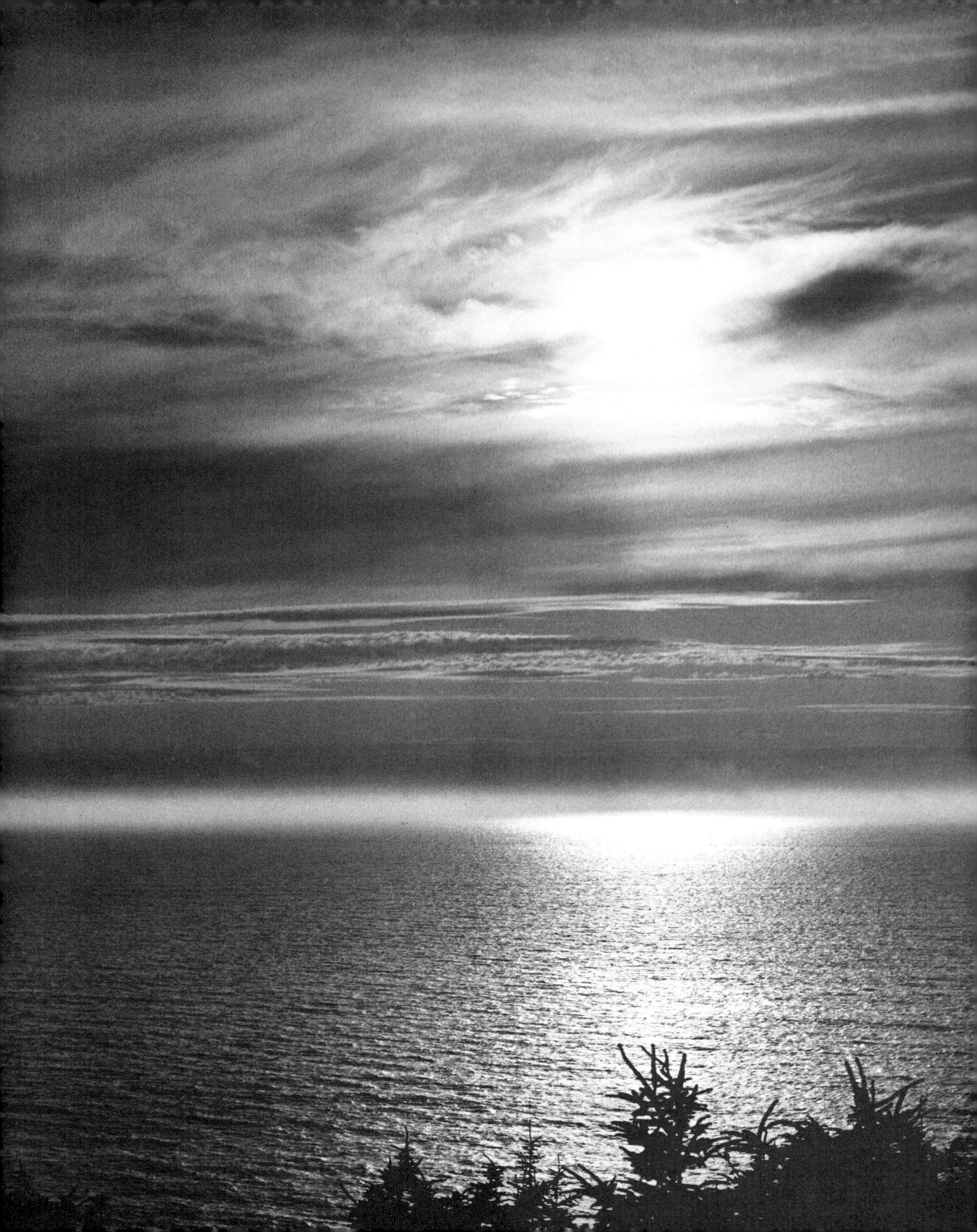

CLARK / November 7th Thursday 1805

... our Small Canoe which got seperated in the fog this morning joined us this evening ....

Great joy in camp   we are in view of the Ocian [actually, Gray's Bay], (in the morning when fog cleared off just below last village (first on leaving this village) of Warkiacum) this great Pacific Octean which we been so long anxious to See. and the roreing or noise made by the waves brakeing on the rockey Shores (as I suppose) may be heard disti[n]ctly

We made 34 miles to day as computed.

CLARK / November 9th Saturday 1805

... at 2 oClock P M the flood tide came in accompanied with emence waves and heavy winds, floated the trees and Drift which was on the point on which we Camped ... maney of [the trees] nearly 200 feet long and from 4 to 7 feet through. our camp entirely under water dureing the hight of the tide ... notwithstanding the disagreeable Situation of our party ... they are chearfull and anxious to See further into the Ocian, The water of the river being too Salt to use we are obliged to make use of rain water. ...

CLARK / November 11th Monday 1805

... the great quantites of rain which has loosened the Stones on the hill Sides; and the Small stones fall down upon us, our canoes at one place at the mercy of the waves, our baggage in another; and our selves and party Scattered on floating logs and Such dry Spots as can be found on the hill sides, and crivicies of the rocks. we purchased of the Indians 13 red charr which we found to be an excellent fish. we have seen those Indians above and are of a nation who reside above and on the opposit Side who call themselves Calt har mar. [Cath lah ma]   they are badly clad & illy made ... one of those men had on a Salors Jacket and Pantiloons.   and made Signs that he got those clothes from the white people who lived below the point &c. those people left us and crossed the river (which is about 5 miles wide at this place) through the highest waves I ever Saw a Small vestles ride. Those Indians are certainly the best Canoe navigaters I ever Saw.   rained all day.

November 7, 1805

CLARK / *November 12th Tuesday 1805*

... our Situation is dangerous. we took the advantage of a low tide and moved our camp around a point to ... this cove .... It would be distressing to See our Situation, all wet and colde our bedding also wet, (and the robes of the party which compose half the bedding is rotten and we are not in a Situation to supply their places) .... Fortunately for us our men are healthy. ...

CLARK / *November 13th Wednesday 1805*

Some intervals of fair weather last night, rain continue[d] this morning. I walked up the Brook & assended the first Spur of the mountain with much fatigue, the distance about 3 miles, through an intolerable thickets of Small pine .... my principal object in assending this mountain was to view the countrey below, the rain continuing and weather proved So cloudy that I could not See any distance. on my return we dispatched 3 men Colter, Willard and Shannon in the Indian canoe to get around the point if possible and examine the river, and the Bay below for a go[o]d harber for our canoes to lie in Safty &c. ... The rain continue all day. nothing to eate but pounded fish which we Keep as a reserve and use in Situations of this kind.

CLARK / *November 14th Thursday 1805*

rained all the last night without intermition, and this morning. wind blows verry hard, but our situation is Such that we cannot tell from what point it comes. one of our canoes is much broken by the waves dashing it against the rocks. 5 Indians came up in a canoe, thro' the waves, which is verry high and role with great fury. They made Signs to us that they saw the 3 men we Sent down yesterday. only 3 of those Indians landed, the other 2 which was women played off in the waves, which induced me to Suspect that they had taken Something from our men below, at this time one of the men Colter return^d by land and informed us that those Indians had taken his Gigg & basket, I called to the Squars to land and give back the gigg, which they would not doe untill a man run with a gun, as if he intended to Shute them when they landed, and Colter got his gig & basket   I then ordered those fellows off, and they verry readily cleared out   they are of the War-ci-a-cum N.

Colter informed us that "it was but a Short distance from where we lay around the point to a butifull Sand beech, which continued for a long ways, that he had found a good harber in the mouth of a creek near 2 Indian Lodges—that he had proceeded in the canoe as far as he could for the waves, the other two men Willard & Shannon had proceeded on down

Capt Lewis concluded to proceed on by land & find if possible the white people the Indians say is below and examine if a Bay is Situated near the mouth of this river as laid down by [George] Vancouver in which we expect, if there is white traders to find them &c.

... The rain &c. which has continued without a longer intermition than 2 hours at a time for ten days past has distroy^d the robes and rotted nearly one half of the fiew clothes the party has, perticularley the leather clothes ....

November 13, 1805

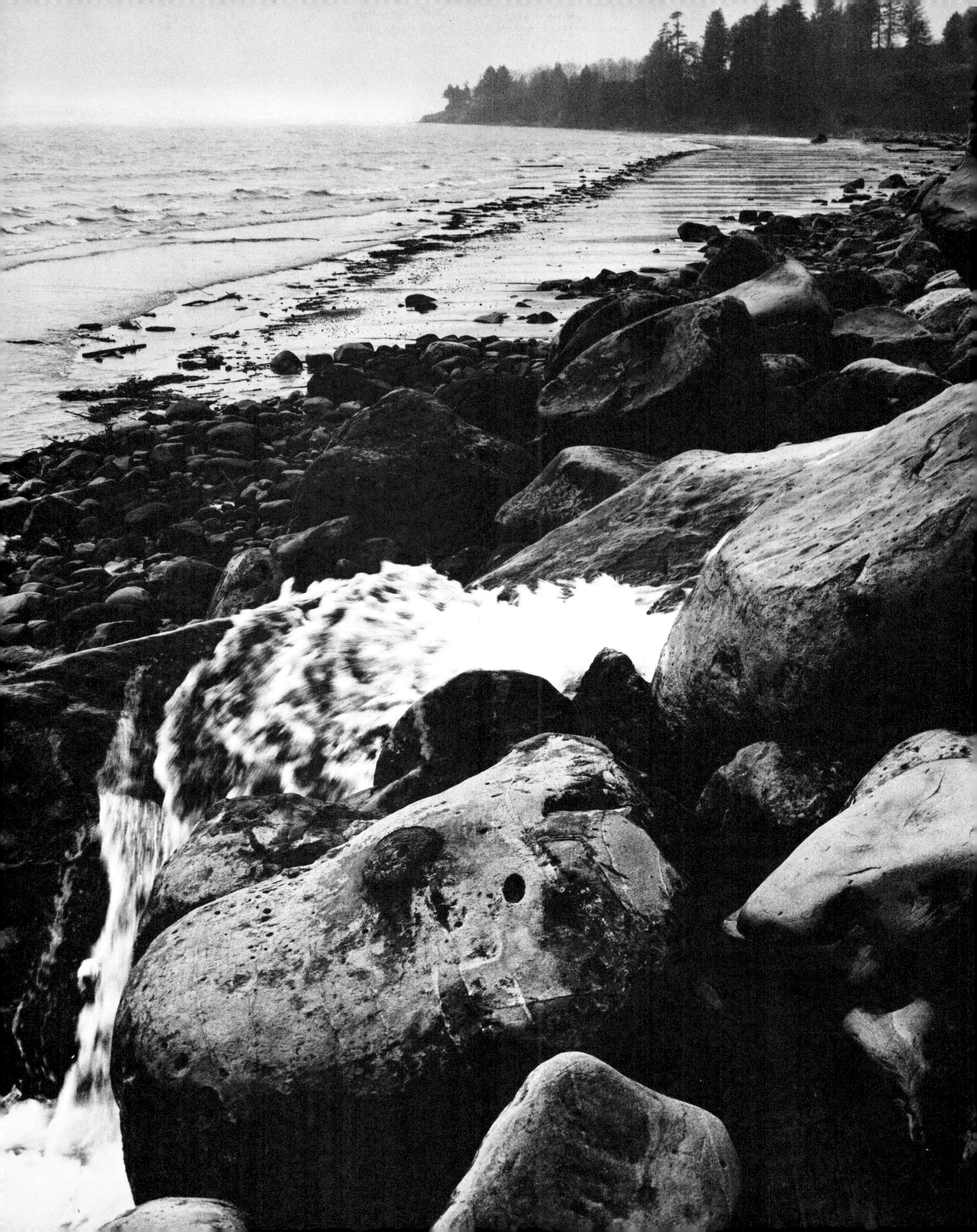

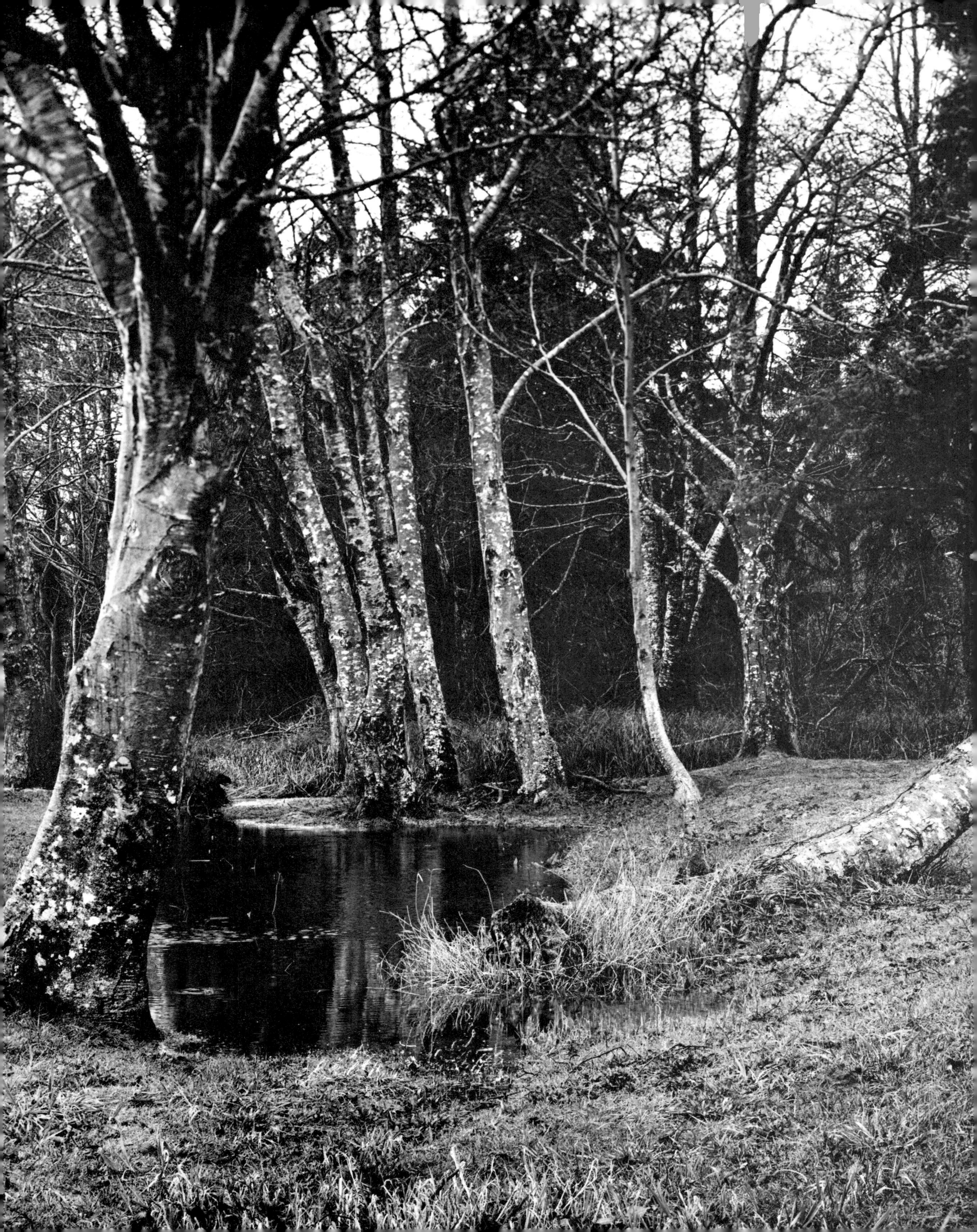

CLARK / November 15*th* Friday 1805

... about 3 oClock the wind luled, and the river became calm, I had the canoes loaded in great haste and Set Out, from this dismal nitich where we have been confined for 6 days passed, without the possibility of proceeding on, returning to a better Situation, or get out to hunt; Scerce of Provisions, and torents of rain poreing on us all the time. proceeded on passed the blustering point [Ellice] below which I found a butifull Sand beech thro which runs a Small river from the hills, below the mouth of this Stream is a village of 36 houses uninhabited by anything except flees, here I met G. Shannon and 5 Indians. ... as the tide was comeing in and the Seas became verry high imediately from the ocian (imediately faceing us) I landed and formed a camp on the highest Spot I could find between the hight of the tides, and the Slashers in a small bottom   this I could plainly See would be the extent of our journey by water .... in full view of the Ocian from Point Adams or Rond ... to Cape Disapointment, I could not see any Island in the mouth of this river as laid down by Vancouver. the Bay which he laies down in the mouth is imediately below me. This Bay we call Haley's [Baker's] bay from a favourite trader with the Indians which they Say comes into this Bay and trades with them   course to Point adams is S.35°.W. about 8 miles   to Cape Disapointment is S.86°W. about 14 miles.... The Indians who accompanied Shannon from the Village below ... Call themselves Chinnooks, I told those people that they had attempted to Steal 2 guns &c. that if any one of their nation stole any thing that the Senten*l* whome they Saw near our baggage with his gun would most certainly Shute them,   they all promised not to tuch a thing, and if any of their womin or bad boys took any thing to return it imediately and chastise them for it. I treated those people with great distance....

CLARK / November 16*th* Saturday 1805

... this morning clear and butifull ....
    The Countrey on the Star*d* Side above Haleys Bay is high broken and thickley timbered   on the Lar*d* Side from Point Adams the countrey appears low for 15 or 20 miles back to the mountains ... as the Opposit [shore] is too far distant to be distinguished well, I shall not attempt to describe any thing on that side at present. our hunters and fowlers killed 2 Deer 1 Crain & 2 Ducks, and my man York killed 2 geese and 8 Brant ....

November 15, 1805

November 16, 1805

November 18, 1805

CLARK / *November 17*th *Sunday 1805*

a fair cool morning wind from the East. The tide rises at this place 8 feet 6 inches and comes in with great waves . . . .

at half past 1 o Clock Capt *Lewis returned haveing travesed Haley Bay to Cape Disapointment and the Sea coast to the north for Some distance. . . . I directed all the men who wished to see more of the main Ocian to prepare themselves to Set out with me early on tomorrow morning. . . .*

CLARK / *November 18!ʰ Monday 1805*

A little cloudy this morning I set out with 10 men and my man York to the Ocian by land. i. e. Serj! Ordway & Pryor, Jos. & Ru Fields, Go. Shannon, W. Brattin, J. Colter, P. Wiser, W. Labieche & P. Shabono one of our interpreters & York. I set out at Day light and proceeded....

N. 80°. W. 1 Mile to a point of rocks about 40 feet high, from the top of which the hill Side is open and assend with a Steep assent to the tops of the mountains....

N. W. 7 Mile[s] to the enterance of a creek at a lodge or cabin of Chinnooks passing on a wide Sand bar the bay to my left and Several Small ponds containing great numbers of water fowls to my right; with a narrow bottom of alder & Small balsam between the Ponds and the Mount!ⁿ at the Cabin I saw 4 womin and Some children one of the women in a desperate Situation, covered with sores scabs & ulsers no doubt the effects of venereal disorders which Several of this nation which I have Seen appears to have.

... S. 46°. E. 2 Miles to the iner extremity of Cape Disapointment passing ... through a low isthmus. this Cape is an ellivated circlier [circular] point covered with thick timber on the iner Side and open grassey exposur next to the Sea and rises with a Steep assent.... I assended this hill which is covered with high corse grass. decended to the N. of it and camped. [walked] 19 Miles [to-day]....

from Cape Disapointment to a high point of a Mount!ⁿ which we shall call [Clarke's Point of View] [False Tillamook Head] beares S. 20° W. about 40 [25] miles, point adams is verry low and is Situated within the derection between those two high points of land, the water appears verry Shole from off the mouth of the river for a great distance, and I cannot assertain the direction of the deepest chanel, the Indians point nearest the opposit Side.... men appear much Satisfied with their trip beholding with estonishment the high waves dashing against the rocks & this emence Ocian

November 18, 1805

Cape Disapointment at the Enterance of the Columbia River into the Great South Sea or Pacific Ocean

CLARK / *Tuesday November the 19th 1805*

... after takeing a Sumptious brackfast of Venison which was rosted on Stiks exposed to the fire, I proceeded on through ruged Country of high hills and Steep hollers on a course from the Cape N. 20° W. 5 miles on a Direct line to the commencement of a Sandy coast which extended N. 10° W. from the top of the hill above the Sand Shore to a Point of high land distant near 20 miles. this point I have taken the Liberty of Calling after my particular friend Lewis. at the commencement of this Sand beech the high lands leave the Sea Coast in a Direction to Chinnook river, and does not touch the Sea Coast again below point Lewis leaveing a low pondey Countrey, maney places open with small ponds in which there is great numb^r of fowl   I am informed that the Chinnook Nation inhabit this low countrey and live in large wood houses on a river which passes through this bottom Parrilal to the Sea coast and falls into the Bay

I proceeded on the sandy coast 4 miles, and marked my name on a Small pine, the Day of the month & year, &^{c} and returned to the foot of the hill, from which place I intended to Strike across to the Bay, I saw a Sturgeon which had been thrown on Shore and left by the tide 10 feet in length, and Several joints of the back bone of a Whale ....

CLARK / *Wednesday November the 20th 1805*

Some rain last night   dispatched Labeech [*Francis Labiche*] to kill some fowl for our brackfast   he returned in about 2 hours with 8 large Ducks on which we brackfast .... I proceeded on up the Beech .... I met Several parties of Chinnooks which I had not before Seen, they were on their return from our Camp.... found maney of the Chin nooks with Capt Lewis of whome there was 2 Cheifs Com com mo ly & Chillar-la-wil to whome we gave Medals and to one a flag.... both Capt Lewis & my self endeavored to purchase the roab with differant articles   at length we precured it for a belt of blue beeds which the Squar-wife of our interpreter Shabono wore around her waste....

November 19, 1805

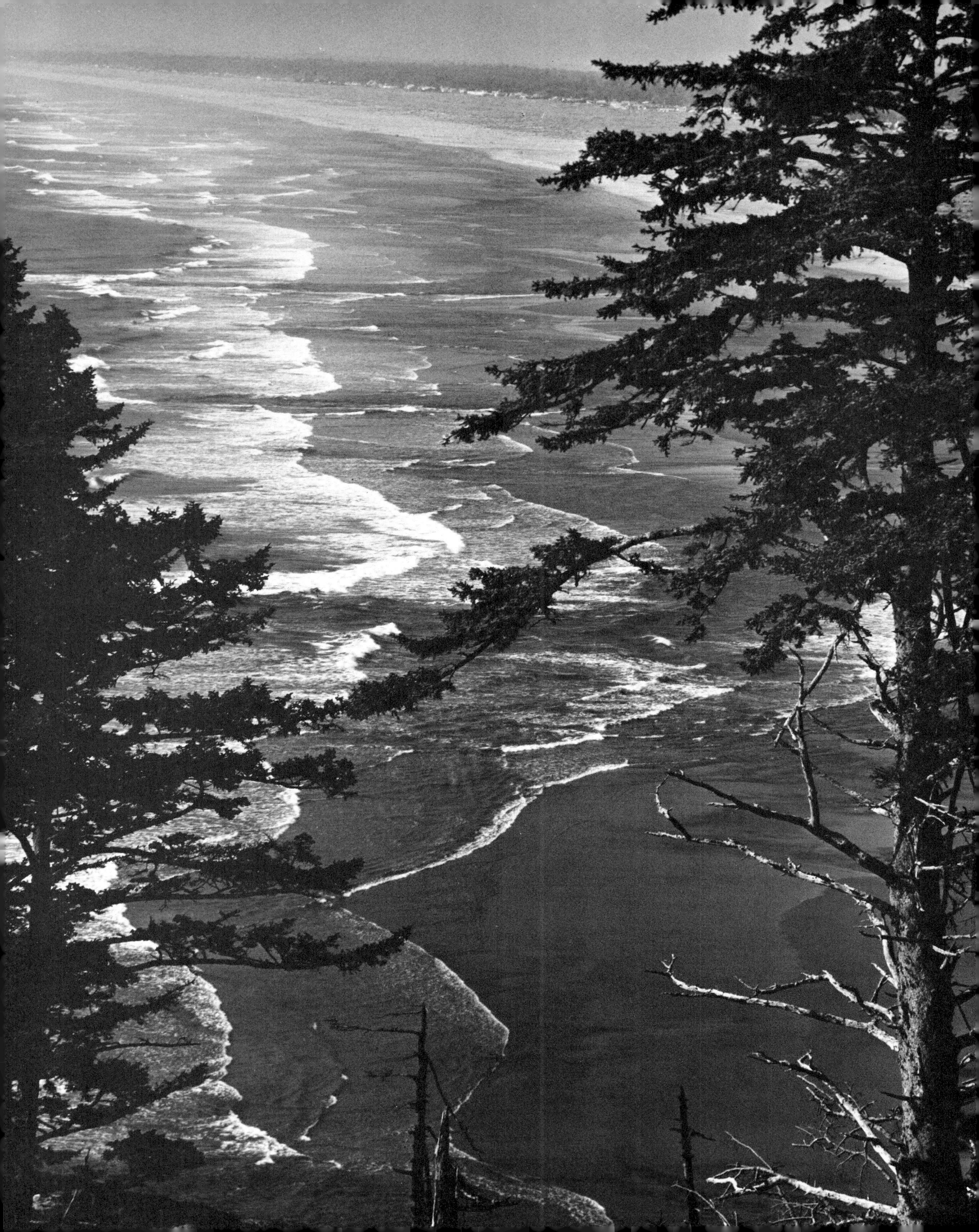

November 21, 1805

CLARK / *Thursday November 21st 1805.*

... *An old woman & Wife to a Cheif of the Chunnooks came and made a Camp near ours. She brought with her 6 young Squars (her daughters & nieces) I believe for the purpose of Gratifying the passions of the men of our party and receving for those indulgiences Such Small [presents] as She (the old woman) thought proper to accept of.*

Those people appear to View Sensuality as a Necessary evel, and do not appear to abhor it as a Crime in the unmarried State.... the Womin of the Chinnook Nation have handsom faces low and badly made with large legs & thighs which are generally Swelled from a Stopage of the circulation in the feet (which are Small) by maney Strands of Beeds or curious Strings which are drawn tight around the leg above the ankle, their legs are also picked [i.e., tattoed] with defferent figures, I saw on the left arm of a Squar the following letters J. Bowman .... The food of this nation is principally fish & roots the fish they precure from the river by the means of nets and gigs, and the Salmon which run up the Small branches together with what they collect drifted up on the Shores of the Sea coast hear to where they live....
[first draft] ... *we divided some ribin between the men of our party to bestow on their favourite Lasses, this plan to save the knives & more valueable articles....*

CLARK / *Friday November 22nd 1805*

a moderate rain all the last night with wind, a little before Day light the wind which was from the S.S.E. blew with Such Violence that we wer almost overwhelmned with water blown from the river, this Storm did not sease at day but blew with nearly equal violence throughout the whole day accompan[i]ed with rain. O! how horriable is the day waves brakeing with great violence against the Shore throwing the Water into our Camp &c. all wet and confind to our Shelters, Several Indian men and women crouding about the mens shelters to day, we purchased a fiew Wappato roots for which we gave Armban[d]s, & rings to the old Squar, those roots are equal to the Irish potato, and is a tolerable substitute for bread ...

CLARK / *Saturday November 22[3]rd 1805.*

A calm Cloudy morning, ... Capt Lewis Branded a tree with his name Date &c. I marked my name the Day & year on a alder tree, the party all Cut the first

November 23, 1805

letters of their names on different trees in the bottom....

 in the evening Seven indians of the Clot sop Nation came over in a Canoe, they brought with them 2 Sea otter Skins for which they asked blue beads &c. and Such high pricies that we were unable to purchase them without reducing our Small Stock of Merchendize, on which we depended for Subcistance on our return up this river. mearly to try the Indian who had one of those Skins, I offered him my Watch, handkerchief a bunch of red beads and a dollar of the American coin, all of which he refused and demanded "ti-â-co-mo-shack" which is Chief beads and the most common blue beads, but fiew of which we have at this time

 This nation is the remains of a large nation destroyed by the Small pox or Some other [disease] which those people were not acquainted with, they Speak the Same language of the Chinnooks and resemble them in every respect except that of Stealing, which we have not cought them at as yet.

November 30, 1805

Clark / Sunday November 24th 1805.

... being now determined to go into Winter quarters as soon as possible, as a convenient Situation to precure the Wild animals of the forest which must be our dependance for Subsisting this Winter, we have every reason to believe that the Nativs have not provisions Suffi[ci]ent for our consumption, and if they had, their prices are So high that it would take ten times as much to purchase their roots & Dried fish as we have in our possession .... Most Elk is on the Opposit Shore, and that the greatest Numbers of Deer is up the river at Some distance .... added to [this] a convenient Situation to the Sea coast where We Could make Salt, and a probility of Vessels comeing into the Mouth of Columbia ("which the Indians inform us would return to trade with them in 3 months") from whome we might precure a fresh Supply of Indian trinkets to purchase provisions on our return home .... to the above advantagies in being near the Sea Coast one most Strikeing one occurs to me i.e., the Climate which must be from every appearance much milder than that above the 1st range of Mountains .... if this Should be the case it will most Certainly be the best Situation of our Naked party dressed as they are altogether in leather.

Clark / November 26th Tuesday 1805

[first draft] ... We are now decending to see if a favourable place should offer on the So Side to winter &c.

Lewis / November 29th 1805

the wind being so high the party were unable to proceed with the perogues. I determined therefore to proceed down the river [Netul River, now Lewis and Clark River] on it's E. side in surch of an eligible place for our winters residence ....

Clark / Saturday 30th of November 1805

Some rain and hail with intervales of fair weather .... Several men Complain of a looseness and griping which I contribute to the diet, pounded fish mixed with Salt water, I derect that in future that the party mix the pounded fish with fresh water. The squar gave me a piece of bread made of flour which She had reserved for her child and carefully Kept untill this time, which has unfortunately got wet, and a little Sour. this bread I eate with great satisfaction, it being the only mouthfull I had tasted for Several months past. my hunters killed three Hawks, which we found fat and delicious, they Saw 3 Elk but could not get a Shot at them. The fowlers killed 3 black Ducks with Sharp White beeks ...

The Chinnooks Cath lâh mâh & others in this neighbourhood bury their dead in their Canoes. ... we cannot understand them Sufficiently to make any enquiries relitive to their religious opinions, from their depositing Various articles with their dead, [they] beleve in a State of future ixistance.

(Copy for Dr. Barton) I walked on the point and observed rose bushes different Species of pine, a Species of ash, alder, a Species of wild Crab Loral. and Several Species of under groth common to this lower part of the Columbia river. The hills on this coast rise high and are thickly covered with lofty pine maney of which are 10 & 12 feet through and more than 200 feet high. ...

Clark / Sunday December 1st 1805.

... The emence Seas and waves ... and this roaring has continued ever Since our arrival in the

December 1, 1805

neighbourhood of the Sea Coast which has been 24 days Since we arrived in Sight of the Great Western; (for I cannot Say Pacific) Ocian as I have not Seen one pacific day Since my arrival in its vicinity, and its waters are forming and petially [perpetually] breake with emenc waves on the Sands and rockey coasts, tempestous and horiable. . . .

CLARK / Tuesday 3d December 1805

. . . I am still unwell and can't eate even the flesh of the Elk. . . . Some rain this evening   I marked my name on a large pine tree imediately on the isthmus   William Clark December 3rd 1805. By Land from the U.States in 1804 & 1805.

CLARK / Wednesday 4th December 1805

. . . my appetite has returned and I feel much better of my late complaint. . . . no account of Capt Lewis. I fear Some accident has taken place in his craft or party.

CLARK / Thursday 5th of December 1805

. . . Capt Lewis returned with 3 men in the Canoe and informs me that he thinks that a Sufficient number of Elk may be pr[o]cured convenient to a Situation on a Small river . . . .

CLARK / Saturday 7th of December 1805

Some rain from 10 to 12 last night,   this morning fair,   have every thing put on board the Canoes and Set out to the place Capt Lewis had viewed and thought well Situated for winter quarters. . . . after brackfast I delayed about half an hour before York Came up, and then proceeded around this Bay which I call [have taken the liberty of calling] Meriwethers Bay [now Young's Bay] . . . we assended a river [Lewis and Clark River] . . . 3 miles to the first point of high land on the West Side . . . this situation is on a rise about 30 feet higher than the high tides leavel and thickly Covered with lofty pine. this is certainly the most eligable Situation for our purposes . . . .

CLARK / Tuesday 10th December 1805

a Cloudy rainey morning   verry early I rose and walked on the Shore of the Sea coast and picked up Several curious Shells.   I saw Indians Walking up and down the beech ... one man came to where I was and ... told me (in English) the "Sturgion

December 10, 1805

December 30, 1805

was verry good" .... after amuseing my self for about an hour on the edge of the rageing Seas I returned to the houses, one of the Indians pointed to a flock of Brant ... I walked down with my Small rifle and killed two at about 40 yds distance, on my return to the houses two Small ducks Set at about 30 Steps from me   the Indians pointed at the ducks they were near together, I Shot at the ducks and accidentally Shot the head of one off ... and every man came around examined the Duck looked at the gun the Size of the ball which was 100 to the pound and Said in their own language ... do not under Stand this kind of Musket &c. .... I proceeded on through a heavy rain to the Camp at our intended fort ... found Capt Lewis with all the men out Cutting down trees for our huts &c. ...

CLARK / Thursday 12.th December 1805

... in the evening two Canoes of Clât Sops Visit us they brought with them Wappato, a black Sweet root ....

Those Indians appear well disposed   we gave a Medal to the principal Chief named Con-ny-au or Com mo-wol .... I can readily discover that they are close deelers, & Stickle for a verry little, never close a bargin except they think they have the advantage   Value Blue beeds highly ....

CLARK / Friday 13.th December 1805

The Clatsops leave us to day .... we Continue to put up the Streight butifull balsom pine on our houses. ...

CLARK / Tuesday 17.th of December 1805

... all the men at work about the houses, Some Chinking Dobbing Cutting out dores &.^c &.^c ...

CLARK / Tuesday 24.th December 1805

hard rain at Different Times last night and all this day without intermition. men all employ.^d in finishing their huts and moveing into them.

Cuscalah ... come up in a canoe with his young brother & 2 Squars ... as we had no files to part with, we each returned the present which we had received, which displeased Cuscalah a little. He then offered a woman to each of us which we also declined axcepting ... the female part appeared to be highly disgusted at our refuseing to axcept of their favours &.^c ...

CLARK / Christmas Wednesday 25.th December 1805

at day light this morning we we[re] awoke by the discharge of the fire arm of all our party & a Selute, Shouts and a Song which the whole party joined in under our windows, after which they retired to their rooms   were chearfull all the morning. after brackfast we divided our Tobacco which amounted to 12 carrots one half of which we gave to the men of the party who used tobacco, and to those who doe not use it we make a present of a handkerchief, The Indians leave us in the evening   all the partly Snugly fixed in their huts. I recved a pres[e]nt of Capt L. of a fleece hosrie Shirt Draws and Socks, a p.^r Mockersons of Whitehouse a Small Indian basket of Gutherich [Silas Goodrich], two Dozen white weazils tails of the Indian woman, & some black root of the Indians before their departure. Drewyer informs me that he saw a Snake pass across the parth to day. The day proved Showerey wet and disagreeable.

we would have Spent this day the nativity of Christ in feasting, had we any thing either to raise our Sperits or even gratify our appetites, our Diner concisted of pore Elk, so much Spoiled that we eate it thro' mear necessity, Some Spoiled pounded fish and a fiew roots.

ORDWAY / Wednesday 25.th Decr 1805.

rainy & wet.   disagreeable weather. We all moved in to our new Fort, which our officers name Fort Clatsop after the name of the Clatsop nation of Indians who live nearest to us. ...

CLARK / Monday 30.th December 1805

... our fortification is completed this evening and at Sun set we let the nativs know that our Custom will be in future, to Shut the gates at Sun Set at which time all Indians must go out of the fort and not return into it untill next morning after Sunrise at which time the gates will be opened, those of the Warciacum Nation who are very fo[r]ward left the houses with reluctianc[e]   this day proved to be the fairest and best which we have had since our arrival at this place, only three Showers dureing this whole day, wind the fore part of the day.

ORDWAY / Monday 30.th Decr 1805.

... we finished puting up our pickets and gates of the fort. ... a centinel placed in the fort to look out for the Savages for our Safety, &C.

January 1, 1806

Fort Clatsop. 1806.
Lewis / January 1st Tuesday. [Wednesday]

This morning I was awoke at an early hour by the discharge of a volley of small arms, which were fired by our party in front of our quarters to usher in the new year . . . . our repast of this day tho' better than that of Christmass, consisted principally in the anticipation of the 1st day of January 1807, when in the bosom of our friends we hope to participate in the mirth and hilarity of the day, and when with the zest given by the recollection of the present, we shall completely, both mentally and corporally, enjoy the repast which the hand of civilization has prepared for us. at present we were content with eating our boiled Elk and wappetoe, and solacing our thirst with our only beverage pure water. . . .

Clark / January 1st 1806

A List of the names of Sundery persons, who visit this part of the Coast for the purpose of trade &c &c in large Vestles; all of which speake the English language &c as the Indians inform us

Moore	Visit them in a large 4 masted ship, they expect him in 2 moons to trade.
I Ey^d [one-eyed] Skellie	in a large ship, long time gorn. . . .
Mr Haley	Visits them in a Ship & they expect him back to trade with them in 3 Moons to trade. he is the favourite of the Indians (from the number of Presents he gives) and has the trade principally with all the tribes.
Washilton	In a Skooner, they expect him in 3 months to return and trade with them—a favourite. . . .
Fallawan	In a Ship with guns he fired on & killed several Indians, he does not trade now . . . .

Lewis / Thursday, January 2nd 1806

. . . we are infested with swarms of flees already in our new habitations; the presumption is therefore strong that we shall not devest ourselves of this intolerably troublesome vermin during our residence here. . . . the fur of both the beaver and otter in this country are extreemly good; those annamals are tolerably plenty near the sea coast, and on the small

Creeks and rivers as high as the grand rappids, but are by no means as much so as on the upper part of the Missouri.

CLARK / Friday the 3rd January 1806

... our party from necescity have been obliged to Subsist some length of time on dogs have now become extreamly fond of their flesh; it is worthey of remark that while we lived principally on the flesh of this animal we wer much more helthy strong and more fleshey then we have been Sence we left the Buffalow Country. as for my own part I have not become reconsiled to the taste of this animal as yet. . . .

LEWIS / Sunday January 5th 1806.

These lads [Alexander Willard and Peter Wiser] also informed us that J. Fields, [William] Bratton and [George] Gibson (the Salt Makers) had with their assistance erected a comfortable camp killed an Elk and several deer and secured a good stock of meat; they commenced the making of salt and found that they could obtain from 3 quarts to a gallon a day; they brought with them a specemine of the salt of about a gallon, we found it excellent, fine, strong, & white; this was a great treat to myself and most of the party, having not had any since the 20th Ult*mo*; I say most of the party, for my friend Cap*t* Clark. declares it to be a mear matter of indifference with him whether he uses it or not; for myself I must confess I felt a considerable inconvenience from the want of it; the want of bread I consider a trivial provided, I get fat meat, for as to the species of meat I am not very particular, the flesh of the dog the horse and the wolf, having from habit become equally formiliar with any other, and I have learned to think that if the chord be sufficiently strong, which binds the soul and boddy together, it dose not so much matter about the materials which compose it. . . .

LEWIS / Monday January 6th 1806.

Capt Clark set out after an early breakfast with the party in two canoes as had been concerted the last evening; Charbono and his Indian woman were also of the party; the Indian woman was very impo[r]tunate to be permited to go, and was therefore indulged; she observed that she had traveled a long way with us to see the great waters, and that now that monstrous fish [whale] was also to be seen, she thought it very hard she could not be permitted to see either (she had never yet been to the Ocean). . . .

ORDWAY / Monday 6th Jan*y* 1806.

Cap*t* Clark and 12 men Set out with one large canoe and the Small one in order to go after Some of the whail on the coast.   about 9 oClock A.M. cleared off pleasant and warm.

CLARK / Tuesday 7th of January 1806

... left Serg*t* gass and one man of my party [Pvt. William] Werner to make Salt and permited Bratten to accompany me . . . . after walking for 2½ miles on the Stones, my guide made a Sudin halt, pointed to the top of the mountain and uttered the word Pe shack which means bad, and made signs that we could not proceed any further on the rocks, but must pass over that mountain . . . and after about 2 hours labour and fatigue we reached the top of this high mountain, from the top of which I looked down with estonishment to behold the hight which we had assended, which appeared to be 10 or 12 hundred feet . . . . [we] proceeded on a bad road and encamped . . . .

January 5, 1806

CLARK / *Wednesday 8th January 1806*

... arived on a butifull Sand Shore on which we continued for 2 miles ... and proceeded to the place the whale had perished .... I returned to the Village of 5 Cabins on ... E co-la or Whale Creek, found the nativs busily engaged boiling the blubber, which they performed in a large Squar wooden trought by means of hot stones .... my utmost exertion ... were not able to precure more blubber than about 300^{lb} and a fiew gallons of oil; Small as this stock is I prise it highly; and thank providence for directing the whale to us; and think him much more to us than he was to jonah, having Sent this Monster to be Swallowed by us in Sted of Swallowing of us as jonah's did.

LEWIS / *Friday (Thursday) January 9th 1806.*

... I have been living for two days past on poor dryed Elk, or jurk as the hunters term it.

The Clatsops Chinnooks &c bury their dead in their canoes.... I cannot understand them sufficiently to make any enquiries relitive to their religeous opinions, but presume from their depositing various articles with their dead that they believe in a state of future existence.

The persons who usually visit the entrance of this river for the purpose of traffic or hunting I believe are either English or Americans; the Indians inform us that they speak the same language with ourselves, and give us proofs of their varacity by repeating many words of English, as musquit, powder, shot [k]nife, file, damned rascal, sun of a bitch &c. whether these traders are from Nootka sound, from some other late establishment on this coast, or immediately from the U'States or Great Brittain, I am at a loss to determine, nor can the Indians inform us....

CLARK / *Friday the 10th of January 1806*

... The nativs in this neighbourhood are excessively fond of Smokeing tobacco. in the act of Smokeing they appear to swallow it as they draw it from the pipe, and for maney draughts together you will not perceive the Smoke they take from the pipe, in the Same manner they inhale it in their longs untill they become Surcharged with the Vapour when they puff it out to a great distance through their norstils and mouth; I have no doubt that tobacco Smoked in this manner becomes much more intoxicating, and that they do possess themselves of all its virtues to the fullest extent; they frequently give us Sounding proofs of its createing a dismorallity of order in the abdomen, nor are those light matters thought indelicate in either Sex....

LEWIS / *Sunday (Saturday) January 11th 1806*

... the Cuthlâhmâhs left us this evening on their way to the Clatsops, to whom they purpose bartering their wappetoe for the blubber and oil of the whale, which the latter purchase for beads &c. from the Killamucks; in this manner there is a trade continually carryed on by the natives of the river each trading some article or other with their neighbours above and below them; and thus articles which are vended by the whites at the entrance of this river, find their way to the most distant nations enhabiting it's waters.

LEWIS / *Tuesday (Monday) January 13th 1806.*

This morning I took all the men who could be spared from the Fort and set out in quest of the flesh of the seven Elk that were killed yesterday, we found it in good order being untouched by the wolves ....

The traders usually arrive in this quarter, as has been before observed, in the month of April, and remain untill October; when here they lay at anchor in a bay within Cape Disappointment on the N. side of the river; here they are visited by the natives in their canoes who run along side and barter their comodities with them, their being no houses or fortification on shore for that purpose....

CLARK / *Tuesday 14th January 1806*

... the nativs inhabiting that noble Stream [the Columbia] (from the enterance of Lewis's river to the neighbourhood of the falls the nativs consume all the fish they Catch either for food or fuel) From Towarnehiooks River or a fiew mil[e]s above the Great falls to the grand rapids inclusive anually prepare about 30,000 lb^s of pounded fish (Chiefly Salmon) for Market, but whether this fish is an article of Commerce with their neighbours or is exclusively Sold to, and Consumed by the nativs of the sea coast, we are at a loss to determine.... the Indians in this neighbourhood ... who prepared this pounded fish made signs that they traded it with people below them for Beeds and trinkets &^{c} and Showed us maney articles of European Manufacture which they obtained for it; the Skillutes and Indians about the great rapids are the intermediate merchants and Carryers ....

CLARK / *Saturday [Thursday] 16th January 1806*

... haveing made up our minds to Stay untill the 1st of April every one appears contented with his Situation, and his fare. it is true we could travel even

January 8, 1806

now on our return as far as the timbered Country reaches, or to the falls of the river, but further it would be madness for us to attempt to proceed untill april, as the indians inform us that the Snows lyes Knee deep in the Columbian Plains dureing the winter .... We Should [not] fo[r]ward our homeward journey any by reaching the Rocky Mountains earlier than the 1st of June which we can effect by Setting out from hence by the 1st of April....

CLARK / Sunday [Friday] 17th January 1806

... it is for the construction of those baskets that Bargrass becoms an article of traffic among the nativs of the Columbia. this grass grows only on their mountains near the Snowey region; the blade is about ⅜ of an inch wide and 2 feet long Smothe plient and strong; the young blades which are white from not being exposed to the Sun or air, are those which are most Commonly employ'd, particularly in their neatest work. ...

ORDWAY / Sunday [January] 19th [1806]

... the men in the fort are employed dressing Elk Skins for Mockasons, &C. ...

CLARK / Wednesday [Monday] 27th January 1806

... Goudrich has recovered from the louis veneri which he contracted from a amorous contact with a chinnook damsel. he was cured as Gibson was with Murcury .... I cannot lern that the Indians have any Simples sovereign specifics in the cure of this disease ....

LEWIS / *Thursday (Wednesday) January 29th 1806.*

Nothing worthy of notice occurred today. our fare is the flesh of lean elk boiled with pure water, and a little salt. the whale blubber which we have used very sparingly is now exhausted. on this food I do not feel strong, but enjoy the most perfect health; a keen appetite supplys in a great degree the want of more luxurious sauses or dishes, and still renders my ordinary meals not uninteresting to me, for I find myself sometimes enquiring of the cook whether dinner or breakfast is ready. . . .

LEWIS / *Sunday February 2ed 1806.*

Not any occurrence today worthy of notice; but all are pleased, that one month of the time which binds us to Fort Clatsop and which seperates us from our friends has now elapsed. one of the games of amusement and wrisk of the Indians of this neighbourhood like that of the Sosones consists in hiding in the hand some small article about the size of a bean . . . if they hit on the ha[n]d which contains the peice they win the wager otherwise loose. the individual who holds the peice is a kind of banker and plays for a time being against all the others in the room; when he has lost all the property which he has to venture, or thinks proper at any time, he transfers the peice to some other who then also becoms banker. The Sosone and Minnetares &c have a game of a singular kind but those divide themselves in two parties and play for a common wager . . . .

LEWIS / *Thursday February 6th 1806.*

. . . A species of fir which one of my men informs me is precisely the same with that called the balsam fir of Canada. it grows here to considerable size, being from 2½ to 4 feet in diameter and rises to the hight of eighty or an hundred feet. it's stem is simple branching, ascending and proliferous. . . .

LEWIS / *Tuesday February 13th 1806.*

. . . the Indians . . . informed us that one More who sometimes touches at this place and trades with the natives of this coast, had on board of his vessel three Cows, and that when he left them he continued his course along the N.W. coast. I think this strong circumstancial proof that there is a stettlement of white persons at Nootka sound or some point to the N.W. of us on the coast. . . .

February 6, 1806

LEWIS / *Friday February 14th 1806.*

. . . on the 11th inst Capt Clark completed a map of the country through which we have been passing from Fort Mandan to this place. . . . we now discover that we have found the most practicable and navigable passage across the Continent of North America . . . . by way of the Missouri (falls of Missouri) to the entrance of Dearborn's river . . . from thence to flathead (Clarks) river (by land to) at the entrance of Traveller's rest Creek to the forks, from whence you pursue a range of mounttains which divides the waters . . . of the Kooskooske river by water to the S.E. branch of the Columbia . . . and with the latter to the Pacific Ocean.

LEWIS / *Saturday February 15th 1806.*

... about 3 P.M. Bratton arrived from the salt works and informed us that Serg! Pryor and party were on their way with Gibson who is so much reduced that he cannot stand alone and that they are obliged to carry him in a litter. ...

... The horse is confined principally to the nations inhabiting the great plains of Columbia extending from Latitude 40.° to 50.° N. and occupying the tract of country lying between the rocky Mountains and a range [Cascade] of Mountains which pass the columbia river about the great falls or from Longitude 116 to 121 West. ... Their horses appear to be of an excellent race; they are lofty eligantly formed active and durable; in short many of them look like the fine English coarsers and would make a figure in any country. ... The natives (except those near the R. Mont) appear to take pains in scelecting their male horses from which they breed, in short those of that discription which I have noticed appeared much the most indifferent. whether the horse was orrigeonally a native of this country or not it is out of my power to determine as we cannot understand the language of the natives . . . . horses are said to be found wild in many parts of this extensive plain country. ... This abundance and cheapness of horses will be extremely advantageous to those who may hereafter attem[p]t the fir trade to the East Indies by way of the Columbia river and the Pacific Ocean. ... Among the Sosones of the upper part of the S. E. fork of the Columbia we saw several horses with spanish brands on them which we supposed had been stolen from the inhabitants of Mexeco.

LEWIS / *Thursday February 20th 1806.*

... Collins ... brought with him some cranberries for the sick. Gibson is on the recovery fast; Bratton has an obstenate cough and pain in his back and still appears to be geting weaker. M?Neal from his inattention to his disorder has become worse.

This forenoon we were visited by Tâh-cum a principal Chief of the Chinnooks and 25 men of his nation. ... as he came on a friendly visit we gave himself and party something to eat and plyed them plentifully with smoke. ... in the evening at sunset we desired them to depart as is our custom and closed our gates. we never suffer parties of such number to remain within the fort all night; for notwithstanding their apparent friendly disposition, their great averice and hope of plunder might induce them to be treacherous. ... we well know, that the treachery of the aborigenes of America and the too great confidence of our countrymen in their sincerity and friendship, has caused the distruction of many hundreds of us. ...

CLARK / *Saturday February 22nd 1806*

We were visited today by two Clatsops women and two boys who brought a parcel of excellent hats made of Cedar bark, and ornemented with bear grass. ... the woodwork and sculpture of these people as well as those hats and the water proof baskits evince an ingenuity by no means common among the Aborigenes of America. ... our sick consisting of Gibson, Bratten, Willard M?Neal and Baptiest La Page is something better   Serj! Ordway is complaining of a coold & headake. we have not had as many sick at one time since we left the settlements of the Illinois. the general complaint appears to be bad colds and fevers, with a violent pain in the head, and back, something I believe of the influenza. ...

February 25, 1806

LEWIS / *Tuesday February 25th 1806*

... The Rackoon is found in the woody country on this coast in considerable quantities. the natives take a few of them in snars and deadfalls; tho' appear not to vallue their skins much, and but seldom prepare them for robes. ...

LEWIS / *Monday March 3rd 1806.*

... no movement of the party today worthy of notice. every thing moves on in the old way and we are counting the days which seperate us from the 1st of April and which bind us to fort Clatsop....

CLARK / *Saturday March 15th 1806*

...We were visited this Afternoon... by De-lash-hel-wilt a Chinnook Chief his wife and six women of his Nation, which the Old Boud[bawd] his wife had brought for Market. this was the same party which had communicated the venereal to several of our party in November last, and of which they have finally recovered. I therefore gave the men a particular charge with respect to them which they promised me to observe....
... The Clams of this coast are very small. the shells consist of two valves which open with a hinge, the shell is smooth thin and of an oval form or like that of the common Muscle and of a skye blue colour; it is of every size under a Inch & 3/4 in length....

LEWIS / *Sunday March 16th 1806.*

Not any occurrence worthy of relation took place today.... two handkerchiefs would now contain all the small articles of merchandize which we possess; the ballance of the stock consists of 6 blue robes one scarlet do one uniform artillerist's coat and hat, five robes made of our large flag, and a few old cloaths trimed with ribbon. on this stock we have wholy to depend for the purchase of horses and such portion of our subsistence from the Indians as it will be in our powers to obtain. a scant dependence indeed, for a tour of the distance of that before us....

LEWIS / *Monday March 17th 1806*

Catel and his family left us this morning. Old Delashelwilt and his women still remain   they have formed a ca[m]p near the fort and seem to be determined to lay close s[i]ege to us but I beleive notwithstanding every effort of their wining graces, the men have preserved their constancy to the vow of celibacy which they made .... we have had our perogues prepared for our departure, and shal set out as soon as the weather will permit. the weather is so precarious that we fear by waiting untill the first of April that we may be detained .... Drewyer returned late this evening from the Cathlahmahs with our canoe which Sergt Pryor had left some days since, and also a canoe which he had purchased from those people. for this canoe he gave my uniform laced coat and nearly half a carrot of tobacco. it seems that nothing excep[t] this coat would induce them to dispose of a canoe which in their mode of traffic is an article of the greatest val[u]e except a wife, with whom it is equal, and is generally given in exchange to the father for his daughter. I think the U'States are indebted to me another Uniform coat for that of which I have disposed on this occasion was but little woarn. we yet want another canoe, and as the Clatsops will not sell us one at a price which we can afford to give we will take one from them in lue of the six Elk which they stole from us in the winter....

CLARK / *Tuesday March 17 [18]th 1806*

... this morning we gave Delashelwilt a certificate of his good deportment &c and also a list of our names .... Those list's of our names we have given to several of the nativs, and also posted up a copy in our room. ... "The Object of this list is, that through the medium of some civilized person who may see the same, it may be made known to the informed world, that the party consisting of the persons whose names are hereunto annexed; and who were sent out by the Government of the United States in May 1804, to explore the interior of the continent of North America,

March 15, 1806

did penetrate the same by way of the Missouri and Columbia rivers, to the discharge of the latter into the Pacific Ocian . . . from whence they departed the . . . day of March 1806 on their return to the United States by the same rout they had come out."

On the back of lists we added a sketch of the continent of the upper branches of the Missouri with those of the Columbia, particularly of its upper S. E. branch or Lewis's River, on which we also delienated the track we had came and that we ment to pursue on our return, when the same happened to vary. There seemes so many chances against our governments ever obtaining a regular report, through the medium of the savages, and the traders of this coast that we decline makeing any. Our party are too small to think of leaveing any of them to return to the Un! States by Sea, particularly as we shall be necessarily devided into two or three parties on our return in order to accomplish the Object we have in View; and at any rate we shall reach the U. States in all humain probabillity much earlier than a man could who must in the event of his being left here depend for his passage to the U. State[s] on the traders of the coast, who may not return immediately to the U. States. or if they should, might probably spend the next summer in tradeing with the nativs before they would set out on their return.

LEWIS / *Wednesday March 19.th 1806.*

. . . The Killamucks, Clatsops, Chinnooks, Cathlahmahs and Wâc-ki-a-cums (Qu: Wack^ms) *resemble each other as well in their persons and dress as in their habits and manners. their complexion is not remarkable, being the usual copper brown of most of the tribes of North America. . . . the most remarkable trait in their physiognomy is the peculiar flatness and width of forehead which they artificially obtain by compressing the head between two boards while in a state of infancy and from which it never afterwards perfectly recovers. this is a custom among all the nations we have met with West of the Rocky mountains. . . . this process seems to be continued longer with their female than their mail children, and neither appear to suffer any pain from the operation. . . . the large or apparently swolen legs particularly observable in the women are obtained in a great measure by tying a cord tight around the ankle. their method of squating or resting themselves on their hams which they seem from habit to prefer to siting, no doubt contributes much to this deformity of the legs by preventing free circulation of the blood. . . . The dress of the women consists of a robe, tissue, and sometimes when the weather is uncommonly cold, a vest. . . . The garment which occupys the waist, and from thence as low as nearly to the knee before and the ham, behind, cannot properly be denominated a petticoat . . . . when the female stands erect to conceal those parts usually covered from formiliar view, but when she stoops or places herself in many other attitudes, this battery of Venus is not altogether impervious to the inquisitive and penetrating eye of the amorite. This tissue is sometimes formed of little twisted cords of the silk-grass knoted at their ends and interwoven . . . .*

The favorite ornament of both sexes are the common coarse blue and white beads . . . . they are also fond of a species of wampum which is furnished them by a trader whom they call Swipton. . . . I think the most disgusting sight I have ever beheld is these dirty naked wenches. The men of these nations partake of much more of the domestic drudgery than I had at first supposed. they collect and prepare all the fuel, make the fires, assist in cleansing and preparing the fish, and always cook for the strangers who visit them. they also build their houses, construct their canoes, and make all their wooden utensils. the peculiar provence of the woman seems to be to collect roots and manufacture various articles which are prepared of rushes, flags, cedar bark, bear grass or waytape. the management of the canoe for various purposes seems to be a duty common to both sexes . . . .

LEWIS / *Thursday March 20th 1806.*

... Altho' we have not fared sumptuously this winter and spring at Fort Clatsop, we have lived quite as comfortably as we had any reason to expect we should; and have accomplished every object which induced our remaining at this place except that of meeting with the traders who visit the entrance of this river. our salt will be very sufficient to last us to the Missouri where we have a stock in store. it would have been very fortunate for us had some of those traders arrived previous to our departure from hence, as we should then have had it in our power to obtain an addition to our stock of merchandize which would have made our homeward bound journey much more comfortable. many of our men are still complaining of being unwell; Willard and Bratton remain weak, principally I beleive for the want of proper food. I expect when we get under way we shall be much more healthy. it has always had that effect on us heretofore. . . .

GASS / *Thursday [March] 20th [1806]*

... I made a calculation of the number of elk and deer killed by the party from the 1st of December 1805, to the 20th of March 1806, which gave 131 elk, and 20 deer. There were a few smaller quadrupeds killed, such as otter and beaver, and one racoon. . . .

CLARK / *Sunday 23rd March 1806*

... the rained seased and it became fair about Meridian, at which time we loaded our canoes & at I P. M. left Fort Clatsop on our homeward bound journey. at this place we had wintered and remained from the 7th of Decr 1805 to this day and have lived as well as we had any right to expect, and we can say that we were never one day without 3 meals of some kind a day . . . .

ORDWAY / *Sunday 23rd March 1806.*

... Soon after we had set out from fort Clatsop we were met by a party of the chinooks, the old baud and hir Six Girls, they had a canoe, a Sea otter Skin dryed fish & hats for Sale. we ... proceeded on thro Meriwethers Bay. . . .

LEWIS / *Wednesday March 26th 1806.*

The wind blew so hard this morning that we delayed untill 8 A.M. . . . soon after we halted for dinner the two Wackiacums who have been pursuing us since yesterday morning with two dogs for sale, arrived. they wish tobacco in exchange for their dogs which we are not disposed to give as our stock is now reduced to a very few carrots. our men who have been accustomed to the use of this article (Tobaco) and to whom we are now obliged to deny the uce of this article appear to suffer much for the want of it. they substitute the bark of the wild crab which they chew; it is very bitter, and they assure me they find it a good substitute for tobacco. the smokers substitute the inner bark of the red willow and the sacacommis. . . .

LEWIS / *Tuesday April 1st 1806.*

This morning early we dispatched Sergt Pryor with two men in a small canoe up quicksand [Sandy] river with orders to proceed as far as he could and return this evening. . . . the Indians who encamped near us last evening continued with us untill about midday. they informed us that the quicksand river . . . only extendes through the Western mountains as far as the S. Western side of mount hood where it takes it's source. . . . we were now convinced that there must be some other considerable river which flowed into the columbia on it's south side below us which we have not yet seen . . . . We were visited by several canoes of natives in the course of the day; most of whom were decending the [Columbia] river with their women and children. they informed us that they resided at the great rapids and that their relations . . . were much streightened at that place for want of food . . . . This information gave us much uneasiness with rispect to our future means of subsistence. above [the] falls or through the plains from thence to the Chopunnish there are no deer Antelope nor Elk on which we can depend for subsistence; their horses are very poor most probably at this season, and if they have no fish their dogs must be in the same situation. under these circumstances there seems to be but a gloomy prospect for subsistence on any terms; we therefore took it into serious consideration what measures we were to pursue on this occasion; it was at once deemed inexpedient to wait the arrival of the salmon as that would detain us so large a portion of the season that it is probable we should not reach the United States before the ice would close the Missouri; or at all events would hazard our horses which we left in charge of the Chopunnish who informed us they intended passing the rocky mountains to the Missouri as early as the season would permit them w[h]ich is as we believe about the begining of May. . . .

CLARK / *Wednesday April 2nd 1806*

... several canoes of the nativs arived at our Camp ... two young men whome they informed us lived at the Falls of a large river [the lower Willamette] which discharges itself into the Columbia .... I deturmined to take a small party and return to this river and examine its size .... at 8 miles passed a village on the South side .... at 3 P. M. I landed at a large double house of the Ne-er-che-ki-oo tribe .... I entered one of the rooms of this house and offered several articles to the nativs in exchange for wappato. they were sulkey and they positively refused to sell any. I had a small pece of port fire match in my pocket, off of which I cut a pece one inch in length & put it into the fire and took out my pocket compas and set myself down on a mat on one side of the fire ... the port fire cought and burned vehemently, which changed the colour of the fire; with the magnit I turned the needle of the compas about very briskly; which astonished and alarmed these nativs and they laid several parsles of wappato at my feet, & begged of me to take out the bad fire; to this I consented; at this moment the match being exhausted was of course extinguished and I put up the magnet &c. this measure alarmed them so much that the womin and children took shelter in their beads and behind the men .... I lit my pipe and gave them smoke, & gave the womin the full amount of the roots .... they appeared somewhat passified and I left them and proceeded on....

April 5, 1806

LEWIS / *Saturday April 5th 1806.*

... The dogwood grows abundantly on the uplands in this neighborhood. it differs from that of the United States in the appearance of it's bark which is much smoother ....

April 1, 1806

LEWIS / *Wednesday April 9th 1806.*

This morning early we commenced the operation of reloading our canoes; at 7 A.M. we departed and ... continued our rout to the Wah-clel-lah Village which is situated on the North side of the river about a mile below the beacon rock; here we halted and took breakfast. ... on our way to this village we passed several beautifull cascades which fell from a great hight over the stupendious rocks which closes the river on both sides nearly, except a small bottom on the South side in which our hunters were encamped. the most remarkable of these casscades falls ... perpendicularly over a solid rock into a narrow bottom of the river on the south side. it is a large creek, situated about 5 miles above our encampment of the last evening. several small streams fall from a much greater hight, and in their decent become a perfect mist which collecting on the rocks below again become visible and decend a second time in the same manner before they reach the base of the rocks. the hills have now become mountains ....

April 9, 1806

Clark / Thursday April 10th 1806

... The South Side of the river is impassable. As we had but one sufficent toe roap and were obliged to employ the cord in getting on our canoes the greater part of the way we could only take them one at a time which retarded our progress ....

Lewis / Friday April 11th 1806.

... a few men were absolutely necessary at any rate to guard our baggage from the War-clel-lars who crouded about our camp in considerable numbers. these are the greates[t] theives and scoundrels we have met with.... two of these fellows met with John Sheilds ... and pushed him out of the road.... I am convinced that no other consideration but our number at this moment protects us.... The salmon have not yet made their appearance ....

Ordway / Monday 14th of April 1806.

... we bought a number of dogs from the natives. they gave us such as they had to eat which was pounded Salmon thistle roots & wild onions ....

Lewis / Monday April 14th 1806.

... we kept close along the N. shore all day. the river from the rapids as high as the commencement of the narrows is from ½ to ¾ of a mile in width, and possesses scarcely any current. the bed is principally rock except at the entrance of Labuish's river which heads in Mount hood and like the quicksand river brings down from thence vast bodies of sand. the mountains through which the river passes nearly to the sepulchre rock, are high broken, rocky, partially covered with fir white cedar, and in many places exhibit very romantic seenes. some handsome cascades are seen on either hand tumbling from the stupendious rocks of the mountains into the river....

Ordway / Wednesday 16th of April 1806.

a clear pleasant morning.... Capt Clark and 8 more of the party went across the River and took Some marchandize & other articles in order to purchase horses &c.... Serg^t Gass and 2 men Set at makeing pack Saddles....

April 9, 1806

April 18, 1806

LEWIS / *Friday April 18th 1806.*

... the long narrows are much more formidable than they were when we decended them last fall   there would be no possibility of passind[g] either up or down them in any vessel. ... I walked up to the Skillute Village and jouined Capt. C. he had procured four horses only ....

CLARK / *Friday 18th April 1806*

... about 10 A. M. the Indians came down from the Eneesher Villages and I expected would take the articles which they had laid by yesterday. but to my estonishment not one would make the exchange to day. two other parcels of goods were laid by, and the horses promised at 2 P.M. I payed but little attention to this bargain, however suffered the bundles to lye. I dressed the sores of the principal Chief gave some small things to his children and promised the chief some Medicine for to cure his sores. his wife who I found to be a sulky Bitch and was somewhat efflicted with pains in her back. this I thought a good oppertunity to get her on my side giveing her something for her back. I rubed a little camphere on her temples and back, and applyed worm flannel to her back which she thought had nearly restored her to her former feelings. this I thought a favourable time to trade with the chief who had more horses than all the nation besides. I accordingly made him an offer which he excepted and sold me two horses. ... I had not slept but very little for the two nights past on account of mice & Virmen with which those indian houses abounded ....

CLARK / *Saturday 19th April 1806.*

We deturmined to make the portage to the head of the long narrows with our baggage and 5 small canoes, the 2 large canoes we could take no further and therefore cut them up for fuel. we had our small canoes drawn up very early and employed all hands in transporting our baggage on their backs and by means of 4 pack horses, over the portage. . . .

The long narrows are much more formidable than they were when we decended them last fall. there would be no possibility of passing either up or down them in any vessle at this time. . . .

LEWIS / *Sunday April 20th 1806.*

. . . I ordered the indians from our camp this evening and informed them that if I caught them attempting to perloin any article from us I would beat them severely. they went off in reather a bad humour and I directed the party to examine their arms and be on their guard. they stole two spoons from us in the course of the day.

CLARK / *Sunday 20th April 1806*

. . . I precured a sketch of the Columbia and its branches of those people in which they made the river which falls into the Columbia imediately above the falls on the South Side to branch out into 3 branches one of which they make head in Mt Jefferson, one in mount Hood and the other in the S W. range of mountains, and does not water that extensive country we have heretofore calculated on. a great portion of the Columbia and Lewis's river and betwen the same and the waters of Callifornia must be watered by the Multnomah [Willamette] river.

GASS / *Monday [April] 21st. [1806]*

. . . made preparations for setting out from this place. While . . . making preparations . . . an Indian stole some iron articles from among the men's hands; which so irritated Captain Lewis, that he struck him; which was the first act of the kind, that had happened during the expedition. The Indians however did not resent it . . . .

LEWIS / *Monday April 21st 1806.*

. . . I detected a fellow in stealing an iron socket of a canoe pole and gave him several severe blows and mad[e] the men kick him out of camp. I now informed the indians that I would shoot the first of them that attempted to steal an article from us. that we were not affraid to fight them, that I had it in my power at that moment to kill them all and set fire to their houses, but it was not my wish to treat them with severity provided they would let my property alone. that I would take their horses if I could find out the persons who had stolen the tommahawks, but that I had reather loose the property altogether than take the ho[r]se of an inosent person. the chiefs [who] were present hung their heads and said nothing.

CLARK / *Monday 21st April 1806*

A fair cold morning   I found it useless to make any further attempts to trade horses with those unfriendly people who only crowded about me to view and make their remarks and smoke, the latter I did not indulge them with to day. at 12 oClock Capt Lewis and party came up from the Skillutes Village with 9 horses packed and one which bratten who was yet too weak to walk, rode, and soon after the two small canoes also loaded with the residue of the baggage which could not be taken on horses. we had every thing imediately taken above the falls. . . .

April 20, 1806

April 23, 1806

LEWIS / *Tuesday April 22ed 1806.*
... Charbono's horse threw his load, and taking fright at the saddle and robe which still adhered, ran at full speed down the hill, near the village he disengaged himself from the saddle and robe, an indian hid the robe in his lodge. ... being now confident that the indians had taken it I sent the Indian woman [Sacajawea] on to request Capt. C. to halt the party and send back some of the men to my assistance being deturmined either to make the indians deliver the robe or birn their houses. they have vexed me in such a manner by such repeated acts of villany that I am quite disposed to treat them with every severyty, their defenseless state pleads forgivness so far as rispects their lives. with this resolution I returned to their village which I had just reached as Labuish met me with the robe .... I now returned and joined Capt. Clark .... from the top of this emmenense Capt. C. had an extensive view of the country. he observed the range of mountains in which Mount Hood stands to continue nearly south as far as the eye could reach. he also observed the snow-clad top of Mount Jefferson which boar S. 10.W.   Mount Hood from the same point boar S. 30. W.   the tops of the range of western mountains are covered with snow. Capt. C. also discovered some timbered country in a Southern direction from him at no great distance. ...

ORDWAY / *Wednesday 23rd of April 1806.*
... towards evening we arived at a large village at the mouth of a creek where we Camped ....

LEWIS / *Wednesday April 23rd 1806.*
... after we had arranged our camp we caused all the old and brave men to set arround and smoke with us. we had the violin played and some of the men danced; after which the natives entertained us with a dance after their method. this dance differed from any I have yet seen. they formed a circle and all sung as well the spectators as the dancers who performed within the circle. these placed their sholders together with their robes tightly drawn about them and danced in a line from side to side, several parties of from 4 to seven will be performing within the circle at the same time. the whole concluded with a premiscuous dance in which most of them sung and danced. these people speak a language very similar to the Chopunnish whome they also resemble in their dress ....

CLARK / *Thursday 24th April 1806*
... we purchased 3 horses and hired 3 others of the Chopunnish man who accompanies us with his family, and at 1 P.M. set out and proceeded on through a open countrey rugid & sandy between some high lands and the river to a village of 5 Lodges of the Met-cow-we band ....

LEWIS / *Thursday April 24th 1806.*
... the natives had tantalized us with an exchange of horses for our canoes in the first instance, but when they found that we had made our arrangements to travel by land they would give us nothing for them I determined to cut them in peices sooner than leave them on those terms, Drewyer struck one of the canoes and split of[f] a small peice with his tommahawk, they discovered us determined on this subject and offered us several strands of beads for each which were accepted. ... many of the natives pased and repassed us today on the road and behaved themselves with distant rispect towards us. most of the party complain of the soarness of their feet and legs this evening; it is no doubt caused by walking over the rough stones and deep sands after b[e]ing for some months passed been accustomed to a soft soil.

LEWIS / *Friday April 25th 1806.*

... *the* Pish-quit-pahs, *may be considered hunters as well as fishermen as they spend the fall and winter months in that occupation. they are generally pleasently featured of good statu[r]e and well proportioned. both women and men ride extreemly well. . . . almost all the horses which I have seen in the possession of the Indians have soar backs. . . . we continued our rout about nine miles where finding as many willows as would answer our purposes for fuel we encamped for the evening. . . . the river hills are about 250 feet high and generally abrupt and craggey in many places faced with a perpendicular and solid rock. this rock is black and hard. leve[l] plains extend themselves from the tops of the river hills to a great distance on either side of the river. . . . I did not see a single horse which could be deemed poor and many of them were as fat as seals. their horses are generally good. . . . had the fiddle played at the request of the natives and some of the men danced. . . .*

LEWIS / *Monday April 28th 1806.*

... *the fiddle was played and the men amused themselves with dancing about an hour. we then requested the [Yakima and Walla Walla] Indians to dance which they very cheerfully complyed with; they continued their dance untill 10 at night. the whole assemblage of indians about 550 men women and children sung and danced at the same time. most of them stood in the same place and merely jumped up to the time of their music. some of the men who were esteemed most brave entered the spase arrond which the main body were formed in solid column, and danced in a circular manner sidewise. at 10 P.M. the dance concluded and the natives retired; they were much gratifyed with seeing some of our party join them in their dance.*

April 25, 1806

LEWIS / *Tuesday April 29th 1806.*

This morning Yellept furnished us with two canoes and we began to transport our baggage over the river . . . . we had now a store of 12 dogs for our voyage through the plains. . . .

The Wallahwollah river discharges itself into the Columbia on it's S. side 15 miles below the entrance of Lewis's river or the S.E. branch. a high range of hills pass the Columbia just below the entrance of this river. . . .

GASS / *Friday [May] 2d. [1806]*

A fine morning. Last night about 9 o'clock, three of the Wal-la-wal-las came up with us, and brought a steel trap that had been left at our camp on the north side of the Columbia . . . . perhaps one of the greatest instances of honesty ever known among Indians. . . .

ORDWAY / *Saturday 3rd May 1806.*

a little rain the later part of last night, and continues Showery and cold   a little hail & Snow intermixed. . . . made 28 miles this day, having nothing to eat bought the only dog the Indians had with them. the air is very cold.

LEWIS / *Monday May 5th 1806.*

Collected our horses and set out at 7 A.M.   at 4½ miles we arrived at the entrance of the Kooskooske, up the N. Eastern side of which we continued our march . . . . we passed an indian man [who] gave Capt. C. a very eligant grey mare for which he requested a phial of eyewater . . . . My friend Capt. C. is their favorite phisician and has already received many applications. in our present situation I think it pardonable to continue this deseption . . . . We take care to give them no article which can possibly injure them. . . . while at dinner an indian fellow verry impertinently threw a poor half starved puppy nearly into my plait by way of derision for our eating dogs and laughed very heartily at his own impertinence; I was so provoked at his insolence that I caught the puppy and th[r]ew it with great violence at him and stru[c]k him in the breast and face, siezed my tomahawk and shewed him by signs if he repeated his insolence I would tommahawk him, the fellow withdrew apparently much mortifyed and I continued my repast on dog without further molestation. . . .

LEWIS / *Tuesday May 6th 1806.*

. . . Capt. C. was busily engaged for several hours this morning in administering eye-water to a croud of applicants. we once more obtained a plentifull meal, much to the comfort of all the party. . . . The river here called Clark's river is that which we have heretofore called the Flathead river, I have thus named it in honour of my worthy friend and fellow traveller Capt. Clark. . . .

LEWIS / *Wednesday May 7th 1806.*

. . . The Spurs of the Rocky Mountains which were in view from the high plain today were perfectly covered with snow. The Indians inform us that the snow is yet so deep on the mountains that we shall not be able to pass them untill the next full moon or about the first of June; others set the time at still a more distant period. this [is] unwelcom intelligence to men confined to a diet of horsebeef and roots, and who are as anxious as we are to return to the fat plains of the Missouri and thence to our native homes. . . .

May 7, 1806

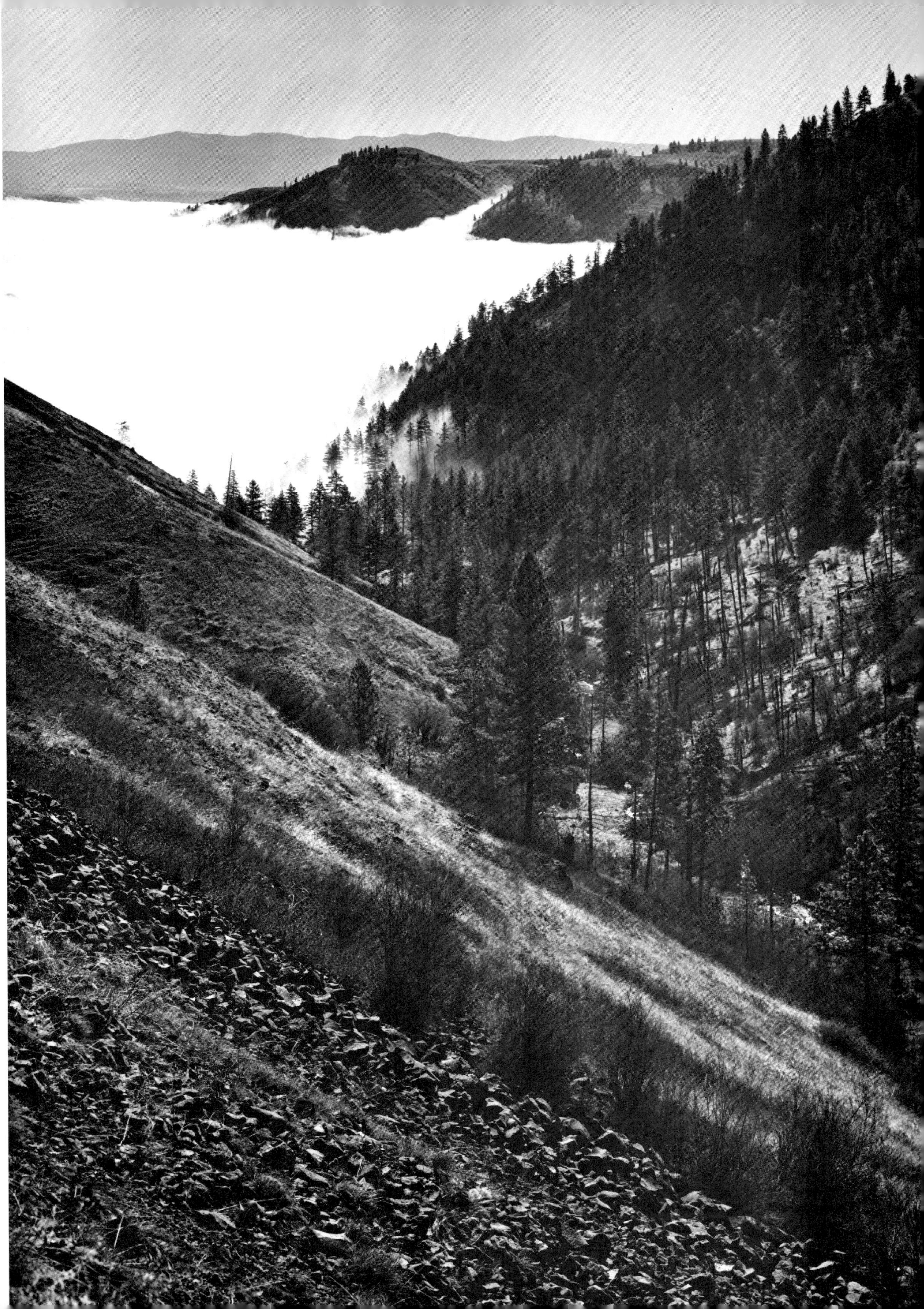

Lewis / Saturday May 10th 1806.

our rout lay through an open plain course S. 35.E. and distance 16 M^s the road was slippery and the snow clogged to the horses feet, and caused them to trip frequently. the mud at the sources of the little ravines was deep black and well supplyed with quawmash. . . . at 4 in the afternoon we decended the hills to Commearp Creek [Lawyer's Canyon Creek] and arrived at the [Nez Percé] Village of Tunnachemootoolt ["Broken Arm"], the cheif at whos lodge we had left the flag last fall. this flag was now displayed on a staff placed at no great distance from the lodge. underneath the flag the Cheif met my friend Capt. C. . . . I came up in a few minutes and we collected the Cheifs and men of consideration smoked with them and stated our situation with rispect to provision. the Cheif spoke to his people and they produced us about 2 bushels of quawmas roots dryed, four cakes of the bread of cows and a dryed salmon trout. we thanked them for this store of provision but informed them that our men not being accustomed to live on roots alone we feared it would make them sick, to obviate which we proposed exchangeing a [good] horse in reather low order for a young horse in tolerable order with a view to kill. the hospitality of the cheif revolted at the eydea of an exchange, he told us that his young men had a great abundance of young horses and . . . they soon produced us two fat young horses one of which we killed, the other we informed them we would pospone killing untill we had consumed the one already killed. This is a much greater act of hospitality than we have witnessed from any nation or tribe since we have passed the Rocky mountains. in short be it spoken to their immortal honor it is the only act which deserves the appellation of hospitallity which we have witnessed in this quarter. . . . The village of the broken arm as I have heretofore termed it consists of one house only which is 150 feet in length built in the usual form of sticks matts and dry grass. it contains twenty four fires and about double that number of families. from appearances I presume they could raise 100 fighting men. the noise of their women pounding roots reminds me of a nail factory. The indians seem well pleased, and I am confident that they are not more so than our men who have their s[t]omachs once more well filled with horsebeef and mush of the bread of cows. the house of coventry is also seen here.

May 10, 1806

Clark / Monday 12th May 1806

a fine morning great numbers of Indians flock about us as usial. . . . Those people are much affraid of the blackfoot indians, and the Big bellies of Fort de prarie . . . . those indians kill great numbers of this nation whenever they pass over to hunt on the Missouri. . . . The Indians brought up a fat horse and requested us to kill and eate it as they had nothing else to offer us to eate. The Cut Nose made a present of a horse to Drewyer at the same time the two horses were offered to Capt Lewis & my self. The horses of those people are large well formed and active. Generally in fine order. sore backs caused by rideing them either with out saddles, or with pads . . . . we gave the twisted hair a gun, powder & 100 ball in part for takeing care of our horses . . . . we have turned our attentions towards the twisted hair who has several sons grown who are well acquainted as well as himself with the various roads through the rocky Mountains and will answer very well as guides to us through those mountains. . . .

Lewis / Monday May 12th 1806.

. . . shot at a mark with the indians, struck the mark with 2 balls dist^{ce} 220 y^{ds}

May 12, 1806

LEWIS / *Tuesday May 13th 1806.*

... The Chopunnish [Nez Percé] are in general stout well formed active men. they have high noses and many of them on the acqueline order with cheerfull and agreeable countenances .... I observed several men among them whom I am convinced if they had shaved their beard instead of extracting it would have been as well supplyed in this particular as any of my countrymen. ... they are expert marksmen and good riders. they do not appear to be so much devoted to baubles as most of the nations we have met with, but seem anxious always to obtain articles of utility, such as knives, axes, tommahawks, kettles blankets and mockersonalls (awls). ... their dress consists of a long shirt which reaches to the middle of [the] thye, long legings which reach as high as the waist, mockersons, and robes. these are formed of various skins and are in all rispects like those particularly discribed of the Shoshones. their women also dress like the Shoshones. ... they also wear a cap or cup on the head formed of beargrass and cedarbark. ...

GASS / *Thursday [May] 15th. [1806]*

... From the Mandan nation to the Pacific ocean, the arms of the Indians are generally bows and arrows, and the war-mallet. The war-mallet is a club with a large head of wood or stone; those of stone are generally covered with leather, and fastened to the end of the club with thongs, or straps of leather, and the sinews of animals. ...

CLARK / *Thursday 15th of May 1806*

... on the high plains off the river the climate is entirely different cool, some snow on the north hill sides near the top and vegetation near 3 weeks later than in the river bottoms, and the rocky Mountains imediately in view covered several say 4 & 5 feet deep with snow. here I behold three different climats within a fiew miles ....

LEWIS / *Saturday May 17th 1806.*

It rained the greater part of the last night and this morning untill 8 OCk the water passed through [the] flimzy covering and wet our bed most perfectly in sho[r]t we lay in the water all the latter part of the night. unfortunately my chronometer which for greater security I have woarn in my fob for ten days past, got wet last night; it seemed a little extraordinary that every part of my breechies which were under my head, should have escaped the moisture except the fob where the time peice was. I opened it and founded

May 17, 1806

May 17, 1806

[it] nearly filled with water which I carefully drained out exposed it to the air and wiped the works as well as I could with dry feathers after which I touched them with a little bears oil. several parts of the iron and steel works were rusted a little which I wiped with all the care in my power. I set her to going and from her apparent motion hope she has sustained no material injury.... it is somewhat astonishing that the grass and a variety of plants which are now from a foot to 18 inches high on these plains sustain no injury from the snow or frost; many of those plants are in blume and appear to be of a tender susceptable texture....

ORDWAY / *Friday 23rd May 1806.*

clear & pleasant.... Wm. bratton having been so long better than 3 months nearly helpless with a Severe pain in his back we now undertake Sweeting him nearly in the manner as the Indians do only cover the hole with blankits having bows bent over the hole. we expect this opperation will help him....

GASS / *Monday [May] 25th. [1806]*

... Our interpreter's child has been very sick, but is getting better....

LEWIS / *Tuesday May 27th 1806*

There is a speceis of Burrowing squirel common in these plains which in their habits somewhat resemble those of the missouri . . . . this little animal measures one fo[o]t five and ½ inches . . . . the appearance of being speckled at a distance. these animals form large ascociations as those of the Missouri . . . . when you approach a burrow the squirrels, one or more, usually set erect on these mounds and make a kind of shrill whistleing nois, something like tweet, tweet, tweet, &c. *they do not live on grass as those of the missouri but on roots. . . .*

ORDWAY / *Wednesday 28th May 1806.*

. . . Saw Several big horn animel or mountain Sheep . . . Some Spots of Snow & falling timber.

May 27, 1806

May 28, 1806

CLARK / *Thursday 29th of May 1806*

... This bird [Clark's nutcracker] feeds on the seeds of the pine and also on insects. it resides in the rocky Mountains at all seasons of the year ....

May 29, 1806

GASS / *Monday [June] 2d [1806].*

... About noon three men, who had gone over to Lewis's river about two and a half days' journey distant, to get some fish, returned with a few very good salmon, and some roots .... One of these men got two Spanish dollars from an Indian for an old razor.—They said they got the dollars from about a Snake Indian's neck they had killed some time ago. There are several dollars among these people, which they get in some way. We suppose the Snake Indians, some of whom do not live very far from New Mexico, get them from the Spaniards in that quarter. The Snake Indians also get horses from the Spaniards. ...

LEWIS / *Sunday June 8th 1806*

... after dark we had the violin played and danced for the amusement of ourselves and the indians. one of the indians informed us that we could not pass the mountains untill the full of the next moon or about the first of July, that if we attemped it sooner our horses would be at least three days travel without food on the top of the mountain; this information is disagreeable .... however as we have no time to loose we will wrisk the chanches ....

Lewis / Monday June 9th 1806

... our party seem much elated with the idea of moving on towards their friends and country, they all seem allirt in their movements today; they have every thing in readiness for a move, and notwithstanding the want of provision have been amusing themselves very merrily today in runing footraces pitching quites [quoits], prison basse &c.... a few days will dry the roads and will also improve the grass.

Clark / Tuesday June 10th 1806.

rose early this morning and had all the horses collected except one of Whitehouses which could not be found .... we packed up and Set out at 11 A M we set out with the party each man being well mounted and a light load on a 2d horse, besides which we have several supernumary horses in case of accident or the want of provisions, we therefore feel ourselves perfectly equiped for the Mountains.... The Country through which we passed is extreemly fertile and generally free from Stone, is well timbered with several Species of fir, long leafed pine and Larch....

Lewis / Wednesday June 11th 1806.

... Whitehouse returned this morning to our camp on the Kooskooske [Clearwater] in surch of his horse. ... [the quamash] root is pallateable but disagrees with me in every shape I have ever used it.

Lewis / Thursday June 12th 1806

... the quawmash is now in blume and from the colour of its bloom at a short distance it resembles lakes of fine clear water, so complete is this deseption that on first sight I could have swoarn it was water.

June 10, 1806

Lewis / Friday June 13th 1806.

... we made a digest of the Indian Nations West of the Rocky Mountains .... they amount by our estimate to 69.000 (about 80,000) Souls.

Lewis / Saturday June 14th 1806.

... we had all our articles packed up and made ready for an early departure .... from hence to traveller's rest we shall make a forsed march .... I am still apprehensive that the snow and the want of food for our horses will prove a serious imbarrassment to us as at least four days journey of our rout in these mountains lies over hights and along a ledge of mountains never intirely destitute of snow. every body seems anxious to be in motion ....

CLARK / Tuesday June 17*th* 1806

... I with great difficulty prosued ... to the top of the mountain where I found the snow from 12 to 15 feet deep ... here was Winter with all it's rigors; the air was cold my hands and feet were benumed. . . . if we proceeded and Should git bewildered in those Mountains the certainty was that we Should lose all of our horses and consequently our baggage enstruments perhaps our papers and thus eventially resque the loss of our discoveries which we had already made if we should be so fortunate as to escape with life. . . . we therefore come to the resolution to return with our horses while they were yet strong and in good order, and indeaver to keep them so untill we could precure an indian to conduct us over the Snowey Mountains . . . .

LEWIS / *Wednesday June 18th 1806.*
... we had not proceeded far this morning before [Pvt. John] Potts ... cut one of the large veigns on the inner side of the leg; I applied a tight bandage with a little cushon of wood and tow on the veign below the wound. Colter's horse fel with him in passing hungry creek and himself and horse were driven down the creek a considerable distance rolling over each other among the rocks. fortunately [he] escaped without injury or the loss of his gun. ... we returned to the glade on the branch of hungry Creek where we had dined on the 16th. inst. ...

LEWIS / *Thursday June 19th 1806.*
... late in the evening Frazier reported that my riding horse that of Capt. Clark and his mule had gone on towards the Quawmash flatts and that he had pursued their tracks on the road about 2-½ miles. ... Cruzatte brought me several large morells which I roasted and eat without salt pepper or grease   in this way I had for the first time the true taist of the morell which is truly an insippid taistless food. our stock of salt is now exhausted except two quarts ....

June 19, 1806

June 19, 1806

June 20, 1806

June 21, 1806

LEWIS / *Friday June 20th 1806.*

... the travelling in the mountains on the snow at present is very good, the snow bears the horses perfictly; it is a firm coa[r]se snow without a crust, and the horses have good foothold without sliping much; the only dificulty is finding the road, and I think the plan we have devised will succeed even should we not be enabled to obtain a guide. ... the snow may be stated on an average at 10 feet deep ....

LEWIS / *Saturday June 21st 1806*

We collected our horses early set out on our return to the flatts. we all felt some mortification in being thus compelled to retrace our steps through this tedious and difficult part of our rout, obstructed with brush and innumerable logs of fallen timber which renders the traveling distressing and even dangerous to our horses. ...

LEWIS / *Tuesday June 24th 1806.*

We collected our horses early this morning and set out accompanyed by our three guides. ... we had fine grass for our horses this evening.

LEWIS / *Wednesday June 25th 1806.*

last evening the indians entertained us with seting the fir trees on fire. they have a great number of dry lims near their bodies which when set on fire creates a very suddon and immence blaze from bottom to top of those tall trees. they are a beatifull object in this situation at night. this exhibition reminded me of a display of fireworks. the natives told us that their object ... was to bring fair weather for our journey....

June 25, 1806

LEWIS / *Thursday June 26th 1806.*

... late in the evening much to the satisfaction of ourselves and the comfort of our horses we ... encamped on the steep side of a mountain ... (having passed a few miles our camp of 18 Sepr 1805). there we found an abundance of fine grass for our horses.... the grass was young and tender of course and had much the appearance of the greensward....

LEWIS / *Friday June 27th 1806.*

... on an elivated point we halted by the request of the Indians a few minutes and smoked the pipe. On this eminence the natives have raised a conic mound of stones of 6 or eight feet high .... from this place we had an extensive view of these stupendous mountains ... we were entirely surrounded by those mountains from which to one unacquainted with them it would have seemed impossible ever to have escaped; in short without the assistance of our guides I doubt much whether we who had once passed them could find our way to Travellers rest in their present

June 27, 1806

June 27, 1806

June 28, 1806

situation for the marked trees on which we had placed considerable reliance are much fewer and more difficult to find than we had apprehended. these fellows are most admireable pilots; we find the road wherever the snow has disappeared though it be only for a few hundred paces . . . . our meat being exhausted we issued a pint of bears oil to a mess which with their boiled roots made an agreeable dish. . . .

GASS / *Friday [June] 27th. [1806]*

. . . The snow is so deep, that we cannot wind along the sides of these steeps, but must slide straight down. The horses generally do [not] sink more than three inches in the snow; but sometimes they break through to their bellies. . . . The day was pleasant throughout; but it appeared to me somewhat extraordinary, to be traveling over snow six or eight feet deep in the latter end of June. . . .

CLARK / *Friday June 27th 1806*

. . . contemplating this Scene Sufficient to have dampened the Sperits of any except such hardy travellers as we have become, we continued our march and at the dist[ance] of 3 M. decended a steep Mountain and passed two small branches of the Chopunnish river just above their fo[r]k, and again assend the ridge on which we passed. at the distance of 7 M. arived at our Encampment of 16th Septr . . .

CLARK / *Saturday June 28th 1806*

. . . the whole of the rout of this day was over deep snows. we find the traveling on the snow not worse than without it, as the easy passage it gives us over rocks and fallen timber fully compensate for the inconvenience of sliping, certain it is that we travel considerably faster on the snow than without it. the snow sinks from 2 to 3 inches with a hors, is coarse and firm . . . .

LEWIS / *Sunday June 29$^{t.h}$ 1806.*

... when we decended from this ridge we bid adieu to the snow. ... at noon we arrived at the quawmas flatts on the Creek of the same name and halted to graize our horses and dine having traveled 12 miles. we passed our encampment of the (13th) September at 10 m^s where we halted there is a pretty little plain of about 50 acres plentifully stocked with quawmash and from apperances this fromes [forms] one of the principal stages or encampments of the indians who pass the mountains on this road. ... after dinner we continued our march seven miles further to the warm springs [Lolo Hot Springs] ... situated ... near the bank of travellers rest creek which at that place is about 10 yards wide. ... the prinsipal spring is about the temperature of the warmest baths used at the hot springs in Virginia. In this bath which had been prepared by the Indians by stoping the run with stone and gravel, I bathed and remained in 19 minutes, it was with dificulty I could remain thus long and it caused a profuse sweat. ... both the men and indians amused themselves with the use of a bath this evening. I observed that the indians after remaining in the hot bath as long as they could bear it ran and plunged themselves into the creek the water of which is now as cold as ice can make it; after remaining here a few minutes they returned again to the warm bath, repeating this transision several times but always ending with the warm bath. I killed a small black pheasant near the quamash grounds this evening ....

LEWIS / *Tuesday July 1st 1806.*

This morning early we sent out all our hunters. set Sheilds at work to repair some of our guns which were out of order [Capt Clark & myself consurted the following plan viz.] *from this place* [Traveller's Rest Creek] *I determined to go with a small party by the most direct rout to the falls of the Missouri, there to leave Thompson McNeal and goodrich to prepare carriages and geer for the purpose of transporting the canoes and baggage over the portage, and myself and six volunteers to ascend Maria's river with a view to explore the country and ascertain whether any branch of that river lies as far north as Lat^d 50. and again return and join the party who are to decend the Missouri, at the entrance of Maria's river. I now called for the volunteers to accompany me on this rout, many turned out, from whom I scelected Drewyer the two Feildes, Werner, Frazier and Serg^t Gass* [accompanied me] *the other part of the men are to proceed with Capt Clark to the head of Jefferson's*

June 29, 1806

July 1, 1806

July 2, 1806

river where we deposited sundry articles and left our canoes. from hence Serg.! Ordway with a party of 9 men are to decend the river with the canoes; Cap.! C. with the remaining ten including Charbono and York will proceed to the Yellowstone river at it's nearest approach to the three forks of the missouri, here he will build a canoe and decend the Yellowstone river with Charbono the indian woman, his servant York and five others to the missouri where should he arrive first he will wait my arrival. Serg.! Pryor with two other men are to proceed with the horses by land to the Mandans and thence to the British posts on the Assinniboin with a letter to M.! Heney (Haney) [Hugh Henney, a Hudson's Bay Company trader] whom we wish to engage to prevail on the Sioux Ch[i]efs to join us on the Missouri, and accompany them with us to the seat of the general government. these arrangements being made the party were informed of our design and prepared themselves accordingly. . . . the indian warrior who overtook us on the 26.th U.lt made me a present of an excellent horse which he said he gave for the good council we had given himself and nation and also to assure us of his attatchment to the white men and his desire to be at peace with the Minnetares of Fort de Prarie. . . .

LEWIS / *Wednesday July 2.cd 1806.*

. . . *I gave the Cheif a medal of the small size; he insisted on exchanging names with me according to their custom which was accordingly done and I was called Yo-me-kol-lick which interpreted is the white bearskin foalded. in the evening the indians run their horses, and we had several foot races . . . . I found several . . . uncommon plant specemines . . . .*

On this date, July 3, 1806, the Expedition divided. Lewis and his party of nine men proceeded from the Bitterroot River through the "short route," Lewis and Clark's Pass to the Falls of the Missouri and an exploration of the headwaters of the Marias River, the north fork of June 8, 1805. His purpose was to determine whether part of the Missouri River drainage might extend past the 49th parallel and thus include some of the profitable Canadian fur trade for the United States.

Clark proceeded to the Yellowstone River and the two parties joined on August 12. Clark's exploration follows Lewis's adventure.

LEWIS / *July 4th 1806*

[first draft] . . . up the buffaloe road river or Co-kah-lah'-ishkit [Big Blackfoot] river. through a timbered country, mountains high rocky and but little bottoms. land poor encamped in a handsom high timbered bottom near the river where there was fine grass . . .

LEWIS / *Friday July 4th 1806.*

[second draft] . . . I gave a shirt a handkercheif and a small quantity of ammunition to the indians. . . . they had cut the meat which I gave them last evening thin and exposed it in the sun to dry . . . . these affectionate people our guides betrayed every emmotion of unfeigned regret at seperating from us; they said that they were confidint that the Pahkees, (the appellation they give the Minnetares) would cut us off. . . .

July 4, 1806

July 4, 1806

July 5, 1806

July 6, 1806

Lewis / *July 5th 1806.*

Set out at 6 A.M....
3 M. to the entrance of a large creek 20 y^ds wide Called Seamans' Creek passing a creek at 1 m. 8 y^ds wide. this course with the river, the road passing through an extensive high prarie rendered very uneven by a vast number of little hillucks and sink-holes....

Lewis / *July 6th 1806.*

Set out a little after sunrise   passed the creek a little above our encampment.
14 M. to the point at which the river leaves the extensive plains and enters the mountains   these plains [Blackfoot Prairie] I called the prarie of the knobs from a number of knobs being irregularly scattered through it....

Lewis / *July 7th 1806.*

... Reubin Fields wounded a moos deer near our camp. my dog much worried....

July 7, 1806

July 7, 1806

July 7, 1806

July 7, 1806

LEWIS / *July 7th 1806.*
... N. 45. E. 2 M. passing the dividing ridge [Lewis and Clark Pass] betwen the waters of the Columbia and Missouri rivers at ¼ of a mile. from this gap which is low and an easy ascent on the W. side of the fort mountain [Square Butte] bears North East, and appears to be distant about 20 Miles....
... N. 20 W 7 Ms over several hills and hollows along the foot of the mountain hights .... saw some sighn of buffaloe early this morning in the valley .... after we encamped Drewyer killed two beaver and shot a third which bit his knee very badly and escaped ...

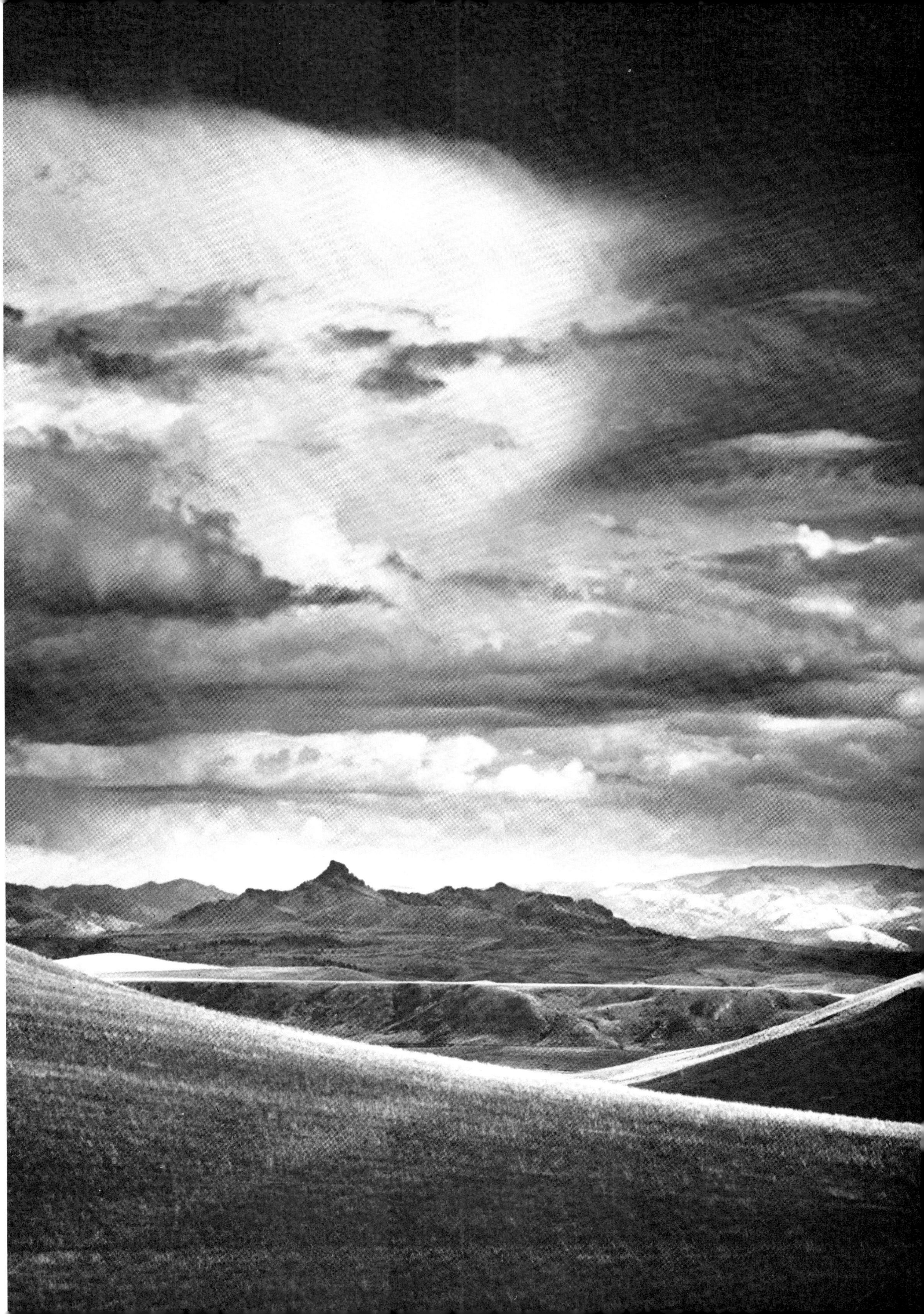

LEWIS / *July 8th 1806.*

... *we steered through the plains leaving the road with a view to strike Medicine [Sun] river and hunt down it to it's mouth in order to procure the necessary skins to make geer, and meat for the three men whom we mean to leave at the falls as none of them are hunters....*

LEWIS / *July 11th 1806.*

the morning was fair and the plains looked beatifull.... the air was pleasant and a vast assemblage of little birds which croud to the groves on the river sung most enchantingly.... it is now the season at which the buffaloe begin to coppelate and the bulls keep a tremendious roaring   we could hear them for many miles.... I sincerely beleif that

July 11, 1806

there were not less than 10 thousand buffaloe within a circle of 2 miles.... set all hands to prepare two canoes the one we made after the mandan fassion with a single [buffalo] skin in the form of a bason [bull-boat]....

LEWIS / *13th July.* [1806]

removed ... to my old station opposite the upper point of the white bear island.... had the cash [cache] opened found my bearskins entirly destroyed by the water.... all my specimens of plants also lost. the Chart of the Missouri fortunately escaped....

LEWIS / *Tuesday July 15th 1806.*

... a little before dark McNeal returned with his musquet broken off at the breach, and informed me that ... he had approached a white bear [grizzly] within ten feet without discover[ing] him the bear being in the thick brush, the horse took the allarm and turning short threw him immediately under the bear; this animal raised himself on his hinder feet for

July 15, 1806

battle, and gave him time to recover from his fall which he did in an instant and with his clubbed musquet he struck the bear over the head ... and broke off the breech, the bear stunned with the stroke fell to the ground and began to scratch his head with his feet; this gave McNeal time to climb a willow tree .... the musquetoes continue to infest us .... my dog even howls with the torture ... they are so numerous that we frequently get them in our thr[o]ats as we breath.

LEWIS / *Wednesday July 16th 1806.*

... proceeded myself with all our baggage and J. Fields down the missouri to the mouth of Medecine river .... these falls have abated much of their grandure since I first arrived at them in June 1805 ... however they are still a sublimely grand object. ...

GASS / *Wednesday [July] 16th. [1806]*

... When Captain Lewis left us, he gave orders that we should wait at the mouth of Maria's river to the 1st of September; at which time, should he not arrive, we were to proceed on and join Captain Clarke at the mouth of the Yellow-stone river, and then to return home; but informed us, that, should his life and health be preserved, he would meet us at the mouth of Maria's river on the 5th of August.

LEWIS / *Thursday July 17th 1806.*

I arrose early this morning and made a drawing of the falls. ... and departed. it being my design to strike Maria's river ... to it's mouth .... the Minnetares of Fort de prarie and the blackfoot indians rove through this quarter of the country and as they are a vicious lawless and reather an abandoned set of wretches I wish to avoid an interview with them if possible. ...

July 17, 1806

LEWIS / Tuesday July 22:d 1806.
... this plain on which we are is very high; the rocky mountains to the S.W. of us appear but low from their base up yet are partially covered with snow nearly to their bases. there is no timber on those mountains within our view; they are very irregular and broken in their form and seem to be composed principally of clay with but little rock or stone. . . . The course of the mountains still continues from S.E. to N.W. . . . I believe that the waters of the Suskashawan apporoach the borders of this river very nearly. I now have lost all hope of the waters of this river ever extending to N. Latitude 50° . . . .

LEWIS / Friday July 25:th 1806.
... I determined that if tomorrow continued cloudy to set out ... with much reluctance without having obtained the necessary data to establish it's longitude . . . .

LEWIS / Saturday July 26:th 1806.
... I had the horses caught and we set out biding a lasting adieu to this place which I now call camp disappointment. I took my rout through the open plains S.E. . . . the country through which this portion of Maria's river passes to the fork which I ascended appears much more broken than that above and between this and the mountains. I had scarcely ascended the hills before I discovered to my left at the distance of a mile an assembleage of about 30 horses, I halted and used my spye glass by the help of which I discovered several indians on the top of an eminence just above them who appeared to be looking down towards the river . . . . this was a very unpleasant sight, however I resolved to make the best of our situation and to approach them in a friendly manner. I directed J. Fields to display the flag which I had brought for that purpose and advanced slowly toward them, about this time they discovered us and appeared to run about in a very confused manner as if much allarmed . . . . I calculated on their number being nearly or quite equal to that of their horses, that our runing would invite pursuit as it would convince them that we were their enimies and our horses were so indifferent that we could not hope to make our escape by flight . . . . under these considerations I still advanced towards them; when we had arrived within a quarter of a mile of them, one of them mounted his horse and rode full speed towards us, which when I discovered I halted and alighted from my horse; he came within a hundred paces halted looked at us and turned his horse about and returned as briskly to his party as he had advanced; while he halted near us I held out my hand and becconed to him to approach but he paid no attention to my overtures. on his return to his party they all decended the hill and mounted their horses and advanced towards us leaving their horses behind them, we also advanced to meet them. . . . and from their known character I expected that we were to have some difficulty with them . . . I should resist to the last extremity prefering death to that of being deprived of my papers instruments and gun and desired that they would

July 22, 1806

July 26, 1806

form the same resolution and be allert and on their guard. when we arrived within a hundred yards of each other the indians except one halted   I directed the two men with me to do the same and advanced singly to meet the indian with whom I shook hands . . . we now all assembled and alighted from our horses; the Indians soon asked to smoke with us, but I told them that the man whom they had seen pass down the river had my pipe and we could not smoke untill he joined us. I requested as they had seen which way he went that they would one of them go with one of my men in surch of him, this they readily concented to and a young man set out with R. Fields in surch of Drewyer. I now asked them by sighns if they were the Minnetares of the North which they answered in the affermative; I asked if there was any chief among them and they pointed out 3   I did not believe them however I thought it best to please them and gave to one a medal to a second a flag and to the third a handkerchief, with which they appeared well satisfyed. . . .

July 26, 1806

July 26, 1806

... I proposed that we should remove to the nearest part of the river and encamp together .... we mounted our horses and rode towards the river which was at but a short distance .... we decended a very steep bluff about 250 feet high to the river where there was a small bottom of nearly ½ a mile in length and about 250 yards wide in the widest part, the river washed the bluffs both above and below us and through it's course in this part is very deep; the bluffs are so steep that there are but few places where they could be ascended, and are broken in several places by deep nitches which extend back from the river several hundred yards, their bluffs being so steep that it is impossible to ascend them. in this bottom there stand t[h]ree solitary trees near one of which the indians formed a large simicircular camp of dressed buffaloe skins and invited us to partake of their shelter which Drewyer and myself accepted .... I learned from them that they were a part of a large [Piegan Blackfoot] band which lay encamped at present near the foot of the rocky mountains .... there was another large band of their nation hunting buffaloe near the broken mountains and were on there way to the mouth of Maria's river .... I told these people that I had come a great way from the East up the large river which runs towards the rising sun, that I had been to the great waters where the sun sets and had seen a great many nations all of whom I had invited to come and trade with me on the rivers on this side of the mountains, that I had found most of them at war with their neighbours and had succeeded in restoring peace among them, that I was now on my way home and had left my party at the falls of the missouri with orders to decend that river to the entrance of Maria's river and there wait my arrival and that I had come in surch of them in order to prevail on them to be at peace with their neighbours particularly those on the West side of the mountains and to engage them to come and trade with me when the establishment is made at the entrance of this river to all which they readily gave their assent and declared it to be their wish to be at peace with the Tushepahs .... I found them extreemly fond of smoking and plyed them with the pipe untill late at night. ... I directed Fields to watch the movements of the indians and if any of them left the camp to awake us all as I apprehended they would attampt to s[t]eal our horses. this being done I feel into a profound sleep and did not wake untill the noise of the men and indians awoke me a little after light in the morning.

July 26, 1806

July 27, 1806

LEWIS / *July 27th 1806. Sunday.*

This morning at daylight the indians got up and crouded around the fire, J. Fields who was on post had carelessly laid his gun down behi[n]d him near where his brother was sleeping, one of the indians the fellow to whom I had given the medal last evening sliped behind him and took his gun and that of his brother unperceived by him, at the same instant two others advanced and seized the guns of Drewyer and myself, J. Fields seeing this turned about to look for his gun and saw the fellow just runing off with her and his brother's he called to his brother who instantly jumped up and pursued the indian with him whom they overtook at the distance of 50 or 60 paces from the camp s[e]ized their guns and rested them from him and R. Fields as he seized his gun stabed the indian [Side Hill Calf] to the heart with his knife the fellow ran about 15 steps and fell dead; of this I did not know untill afterwards, having recovered their guns they ran back instantly to the camp; Drewyer who was awake saw the indian take hold of his gun and instantly jumped up and s[e]ized her and rested her from him but the indian still retained his pouch, his jumping up and crying damn you let go my gun awakened me   I jumped up and asked what was the matter which I quickly learned when I saw drewyer in a scuffle with the indian for his gun. I reached to seize my gun but found her gone, I then drew a pistol from my holster and terning myself about saw the indian making off with my gun   I ran at him with my pistol and bid him lay down my gun which he was in the act of doing when the Fieldses returned and drew up their guns to shoot him which I forbid as he did not appear to be about to make any resistance or commit any offensive act, he droped the gun and walked slowly off, I picked her up instantly . . . as soon as they found us all in possession of our arms they ran and indeavored to drive off all the horses   I now hollowed to the men and told them to fire on them if they attempted to drive off our horses, they accordingly pursued the main party who were dr[i]ving the horses up the river and I pursued the man who had taken my gun who with another was driving off a part of the horses which were to the left of the camp. I pursued them so closely that they could not take twelve of their own horses but continued to drive one of mine with some others; at the distance of three hundred paces they entered one of those steep nitches in the bluff with the horses before them   being nearly out of breath I could pursue no further, I called to them as I had done several times before that I would shoot them if they did not give me my horse and raised my gun, one of them jumped behind a rock and spoke to the other who turned around and stoped at the distance of 30 steps from me and I shot him through the belly, he fell to his knees and on his wright elbow from which position he partly raised himself up and fired at me, and turning himself about crawled in behind a rock which was a few feet from him. he overshot me, being bearheaded I felt the wind of his bullet very distinctly. not having my shotpouch I could not reload my peice and as there were two of them behind good shelters from me I did not think it prudent to rush on them with my pistol which had I discharged I had not the means of reloading untill I reached camp; I therefore returned leasurely towards camp . . . .

July 27, 1806

... I desired him [Drewyer] to haisten to the camp with me and assist in catching as many of the indian horses as were necessary .... we had caught and saddled the horses and began to arrange the packs when the Fieldses returned with four of our horses; we left one of our horses and took four of the best of those of the indian's .... I also retook the flagg but left the medal about the neck of the dead man that they might be informed who we were. we took some of their buffaloe meat and set out ascending the bluffs by the same rout we had decended .... the Fieldses told me that three of the indians whom they pursued swam the river one of them on my horse. and that two others ascended the hill and escaped from them with a part of their horses, two I had pursued into the nitch one lay dead near the camp and the eighth we could not account for but suppose that he ran off early in the contest. having ascended the hill we took our course through a beatifull level plain a little to the S. of East. my design was to hasten to the entrance of Maria's river as quick as possible .... by dark we had traveled about 17 miles further, we now halted to rest ourselves and horses about 2 hours, we killed a buffaloe cow and took a small quantity of the meat. after refreshing ourselves we again set out by moonlight and traveled leasurely, heavy thunderclouds lowered arround us on every quarter but that from which the moon gave us light. we continued to pass immence herds of buffaloe all night as we had done in the latter part of the day. We traveled untill 2 OCk in the morning having come by my estimate after dark about 20 m^s we now turned out our horses and laid ourselves down to rest in the plain very much fatiegued as may be readily conceived....

Lewis / *July 28th 1806. Monday.*
... as day appeared, I awaked the men and directed the horses to be saddled, I was so soar from my ride yesterday that I could scarcely stand, and the men complained of being in a similar situation ... we again resumed our march .... we had proceeded about 12 miles on an East course when we found ourselves near the missouri ... about 8 miles further ... we heared the report of several rifles very distinctly on the river to our right, we quickly repared to this joyfull sound and had the unspeakable satisfaction to see our canoes coming down. ... I now learned that they had brought all things safe having sustaned no loss nor met with any accident of importance. ... we now reimbarked on board the white perog[u]e and five small canoes and decended the river ....

July 27, 1806

LEWIS / *Thursday August 7th 1806.*
... we set out early resolving if possible to reach the Yelowstone river today which was at the distance of 83 m^s from our encampment of the last evening.... at 4 P. M. we arrived at the entrance of the Yellowstone river [into the Missouri]. I landed at the point and found that Capt Clark had been encamped at this place and from appearances had left it about 7 or 8 days.

GASS / *Thursday [August] 7th. [1806]*
... We discovered ... a few words written or traced in the sand, which were "W.C. a few miles farther down on the right-hand side." ... At night we encamped, after coming above 100 miles....

LEWIS / *Monday August 11th 1806.*

... jus[t] opposite to the birnt hills there happened to be a herd of Elk on a thick willow bar and finding that my observation was lost for the present I determined to land and kill some of them accordingly we put too and I went out with Cruzatte only. we fired on the Elk I killed one and he wounded another, we reloaded our guns and took different routs through the thick willows in pursuit of the Elk; I was in the act of firing on the Elk a second time when a ball struck my left thye about an inch below my hip joint, missing the bone it passed through the left thye and cut the thickness of the bullet across the hinder part of the right thye; the stroke was very severe; I instantly supposed that Cruzatte had shot me in mistake for an Elk as I was dressed in brown leather and he cannot see very well; under this impression I called out to him damn you, you shot me, and looked towards the place from whence the ball had come, seeing nothing I called Cruzatte several times as loud as I could but received no answer. I was now preswaded that it was an indian that had shot me as the report of the gun did not appear to be more than 40 paces from me and Cruzatte appeared to be out of hearing of me; in this situation not knowing how many Indians there might be concealed in the bushes I thought best to make good my retreat to the perogue, calling out as I ran for the first hundred paces as loud as I could to retreat that there were indians hoping to allarm him in time to make his escape also; I still retained the charge in my gun which I was about to discharge at the moment the ball struck me. when I arrived in sight of the perogue I called the men to their arms to which they flew in an instant, I told them that I was wounded but I hoped not mortally, by an indian I beleived and directed them to follow me that I would return & give them battle and releive Cruzatte if possible who I feared had fallen into their hands; the men followed me as they were bid and I returned about a hundred paces when my wounds became so painfull and my thye so stiff that I could scarcely get on; in short I was compelled to halt and ordered the men to proceed and if they found themselves overpowered by numbers to retreat in order keeping up a fire. I now got back to the perogue as well as I could and prepared my self with a pistol my rifle and air-gun being determined as a retreat was impracticable to sell my life as deerly as possible. in this state of anxiety and suspense I remained about 20 minutes when the party returned with Cruzatte and reported that there were no indians nor the appearance of any; Cruzatte seemed much allarmed and declared if he had shot me it was not his intention, that he had shot an Elk in the willows after he left or seperated from me. I asked him whether he did not hear me when I called to him so frequently which he absolutely denied. I do not beleive that the fellow did it intentionally but after finding that he had shot me was anxious to conceal his knowledge of having done so. the ball had lodged in my breeches which I knew to be the ball of the short rifles such as that he had, and there being no person out with me but him and no indians that we could discover I have no doubt in my own mind of his having shot me. with the assistance of Serg^t Gass I took off my cloaths and dressed my wounds myself as well as I could, introducing tents of patent lint into the ball holes, the wounds blead considerably but I was hapy to find that it had touched neither bone nor artery. I sent the men to dress the two Elk which Cruzatte and myself had killed which they did in a few minutes and brought the meat to the river. the small canoes came up shortly with the flesh of one Elk. my wounds being so situated that I could not without infinite pain make an observation I determined to relinquish it and proceeded on. . . . as it was painfull to me to be removed I slept on board the perogue; the pain I experienced excited a high fever and I had a very uncomfortable night. at 4 P. M. we passed an encampment which had been evacuated this morning by Capt. Clark, here I found a note from Capt. C. informing me that he had left a letter for me at the entrance of the Yelow stone river, but that Serg^t Pryor who had passed that place since he left it had taken the letter; that Serg^t Pryor having been robed of all his horses had decended the Yelowstone river in skin canoes and had overtaken him at this encampment. . . .

ORDWAY / *Monday 11th August 1806.*

... about 12 oClock Capt Lewis halted at a bottom on S. Side to kill Some Elk. . . . Capt Lewis killed one and cruzatte killed two, and as he still kept fireing one of his balls hit Capt Lewis in his back side. . . . he instantly called to peter [Cruzatte] but Peter not answering he Suppos^d it to be Indians and run to the canoes and ordered the men to their armes. . . .

LEWIS / *Tuesday August 12th 1806.*

Being anxious to overtake Capt. Clark who from appearance of his camps could be at no great distance before me, we set out early and proceeded with all possible expedition   at 8 A. M. the bowsman informed me that there was a canoe and a camp he beleived of whitemen on the N.E. shore. I directed the perogue and the canoes to come too at this place and found it to be the camp of two hunters from the Illinois by the name of Joseph Dickson and Forest Hancock. these men informed me that Capt. C. had passed them about noon the day before. . . . I remained with these men an hour and a half when I took leave of them and proceeded. while I halted with these men Colter and Collins who seperated from us on the 3rd i[n]st rejoined us. they were well no accedent having happened. . . . my wounds felt very stiff and soar this morning but gave me no considerable pain. there was much less inflamation than I had reason to apprehend there would be. I had last evening applyed a poltice of peruvian barks. at 1 P. M. I overtook Capt. Clark and party and had the pleasure of finding them all well. as wrighting in my present situation is extreemly painfull to me I shall desist untill I recover and leave to my fri[e]nd Capt. C. the continuation of our journal. . . .

Here ends Lewis's trip and begins Clark's exploration of the Yellowstone River. On July 3, when the two parties had separated, Clark went upstream along the Bitterroot River with twenty men, the squaw Sacajawea and her papoose, and "49 horses and a colt." They crossed the Continental Divide through Gibbon's Pass and proceeded to the Three Forks, then to the Yellowstone and eastward by canoe 800 miles to its entrance into the Missouri and the reunion with Lewis on August 12, 1806.

July 3, 1806

July 10, 1806

Ordway / Thursday 3rd 1806, July.

we got up our horses and boath parties Set out about one time. Cap^{ts} Lewis & Clark parted here with their parties and procee^d on   I with Capt Clark up the flat head [Bitterroot] River.... we are now on our way to the head of the Missourie....

Clark / Friday July 4th 1806

...This being the day of the decleration of Independence of the United States and a Day commonly scelebrated by my Country I had every disposition to selebrate this day and therefore halted early and partook of a Sumptious Dinner of a fat Saddle of Venison and Mush of Cows (roots).... we made 30 M^s to day....

Clark / Tuesday 8th July 1806

...after dinner we proceeded... to our encampment of 17 Aug^t [1805] at which place we Sunk our Canoes & buried some articles... the most of the Party with me being Chewers of Tobacco become so impatient to be chewing it that they scercely gave themselves time to take their saddles off their horses before they were off to the deposit. I found every article safe.... The road which we have traveled from travellers rest Creek to this place (this place is the head of Jeffer river where we left our canoes) [is] an excellent road.... The distance is 164 Miles....

Clark / Thursday 10th July 1806

...Serg^t Ordway informed me that the party with him had come on very well, and he thought the canoes could go as farst as the horses.... I deturmined to put all the baggage &^{c} which I intend takeing with me to the river Rochejhone in the canoes and proceed on... to the 3 forks or Madisons & galletens rivers.... in passing down in the course of this day we saw great numbers of beaver.... I saw several large rattle snakes in passing the rattle Snake Mountain they were fierce.

Clark / Sunday 13th July 1806

...after dinner the 6 canoes and the party of 10 men under the direction of Serg^t Ordway set out.... I observe Several leading roads... about 18 or 20 miles distant. The indian woman who has been of great service to me as a pilot through this country recommends a gap in the mountain more south which I shall cross.

Clark / Thursday July 3rd 1806.

We colected our horses and after brackfast I took My leave of Capt. Lewis and the indians and at 8 A M Set out with [20] Men interpreter Shabono & his wife & child.... we proceeded on through the Vally of Clarks river... which we found more boutifully versified with small open plains covered with a great variety of Sweet cented plants, flowers & grass.... Some Snow is also to be Seen on the high points and hollows of the Mountains.... we encamped on the N. Side of a large Creek where we found tolerable food for our horses.... Musquetors very troublesom. one man Jo: Potts very unwell this evening owing to rideing a hard trotting horse; I give him a pill of Opiom which soon releve[d] him.

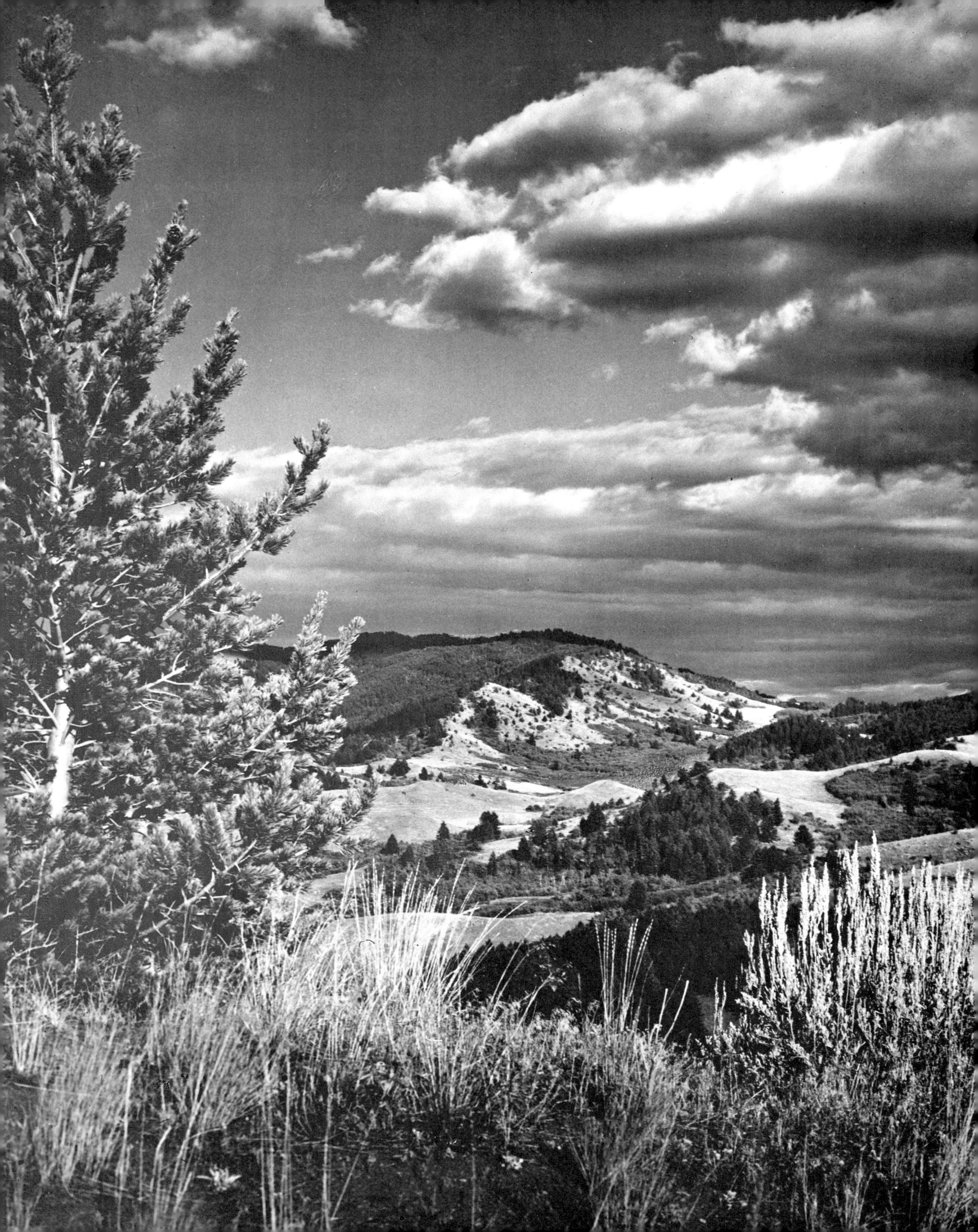

July 14, 1806

Clark / Monday 14th July 1806

... camped on a small branch of the middle fork [Gallatin River] on the NE. side at the commencement of the gap [Bozeman Pass] of the mountain—the road leading up this branch, several other roads all old come in from the right & left. emence quantities of beaver on this Fork ... and their dams very much impeed the navigation of it from the 3 forks down ....

Clark / Tuesday 15th July 1806

We collected our horses and after an early brackf:t at 8 A M set out and proceeded ... over a low gap in the mountain thence across the heads of the N E. branch of the (i.e., the Easterly) fork of Gallitins river which we camped near last night passing over a low dividing ridge to the head of a water course which runs into the Rochejhone, prosueing an old buffalow road which enlargenes by one which joins it from the most Easterly (Northerly) branch of the East fork of Galetine R. proceeding down the branch a little to the N. of East keeping on the North Side of the branch to the River rochejhone at which place I arrived at 2 P M. The distance from the three forks of the Easterly fork of Galletines river (from whence it may be navigated down with small canoes) to the river Rochejhone is 18 Miles on an excellent high dry firm road with very inco[nsi]derable hills. ... after the usial delay of 3 hours to give the horses time to feed and rest and allowing ourselves time also to cook and eate Dinner, I proceeded on down the river ... Great numbers of beaver ....

Clark / Thursday 17th July 1806

... passed ... an Indian fort .... built of logs and bark. ... this work is about 50 feet Diameter & nearly round. the Squaw informs me that when the war parties (of Minnit. Crows &c, who fight Shoshonees) find themselves pursued they make those forts to defend themselves ....

Clark / Saturday 19th July 1806.

... emence Sworms of Grass hoppers *have distroyed every sprig of Grass for maney miles on this side of the river* .... Shabono informed me that he Saw an Indian on the high lands on the opposit side of the river ....

July 15, 1806

July 20, 1806

CLARK / Sunday 20th July 1806

... I deturmined to have two canoes made out of the largest of those trees and lash them together which will cause them to be Study and fully sufficient to take my small party & Self with what little baggage we have down this river. had handles put in the 3 axes and after Sharpening them with a file fell[ed] the two trees which I intended for the two canoes, those trees appeared tolerably Sound and will make canoes of 28 feet in length and about 16 or 18 inches deep and from 16 to 24 inches wide. ...

CLARK / Wednesday 23rd July 1806

... Sgt pryor found an Indian Mockerson and a small piece of roab .... those Indian Signs is conclusive with me that they have taken the 24 horses which we lost on the night of the 20th instant .... I gave Sergt Pryor his instructions and a letter to Mr. Haney ... to engage Mr. Heney to provale on some of the best informed and most influential Chiefs of the different bands of Sieoux to accompany us to the Seat of our Government with a view to let them see our population and resources &c which I believe is the Surest garantee of Savage fidelity to any nation that of a Governmt possessing the power of punishing promptly every aggression. ...

CLARK / Thursday 24th July 1806.

... (good place for fort &c   here the beaver country begins ....) ... I proceeded on [down] the [Yellowstone] river much better than above the enterance of the Clarks fork deep (more navigable) and the Current regularly rapid from 2 to 300 yards in width where it is all together, much divided by islands maney of which are large and well Supplyed with cotton wood trees, some of them large .... for me to mention or give an estimate of the differant Species of wild animals on this particularly Buffalow, Elk Antelopes & Wolves would be increditable. I shall therefore be silent on the subject further. So it is we have a great abundance of the best of meat. we made 70 Ms to day current rapid and much divided by islands. ...

July 24, 1806

July 24, 1806

July 25, 1806

July 25, 1806

CLARK / Friday 25th July 1806.

... *I proceeded on after the* (rain) *lay a little and at 4 P M arived at a remarkable rock situated in an extensive bottom on the Stard Side of the river & 250 paces from it. this rock I ascended and from it's top had a most extensive view in every direction. This rock which I shall call Pompy's Tower is 200 feet high and 400 paces in secumpherance and only axcessable on one Side .... The nativs have ingraved on the face of this rock the figures of animals &c near which I marked my name and the day of the month & year. From the top of this Tower I could discover two low Mountains & the Rocky Mts covered with Snow S W. ...*

CLARK / Sunday 27th July 1806

... *when we pass the Big horn I take my leave of the View of the tremendious chain of Rocky Mountains white with Snow in View of which I have been since the 1st of May last. ...*

CLARK / Tuesday 29th July 1806

... *below this river* [Tongue River] ... *at a fiew Miles from the Rochejhone the hills are high and ruged containing Coal in great quantities. Beaver is very plenty in this part of the Rochejhone. ...*

CLARK / Sunday 1st of August 1806.

... *My situation a verry disagreeable one. in an open canoe wet and without the possibility of keeping my self dry. the country through which we passed is in every respect like that through which I passed yesterday. ... at 2 P. M. I was obliged to land to let the Buffalow cross over. not withstanding an island of a half a mile in width over which this gangue of Buffalow had to pass and the chanel of the river on each side nearly ¼ of a mile in width .... was crowded with those animals for ½ an hour. (I was obliged to lay to for one hour) the other Side of the island for more than ¾ of an hour. ...*

August 3, 1806

August 5, 1806

CLARK / *Tuesday August 3rd 1806*

... at 8. A.M. I arived at the junction of the Rochejhone with the Missouri, and formed my camp imediately in the point between the two river[s] at which place the party had all encamped the 26th of April, 1805....

The distance from the Rocky Mountains at which place I struck the River Rochejhone to its enterance into the Missouri 837 Miles 636 Miles of this distance I decended in 2 Small Canoes lashed together.... The Rochejhone or Yellow Stone River is large and navagable with but fiew obstructions quite into the rocky Mountains....

The colour of the Water differs from that of the Missouri it being of a yellowish brown, whilst that of the Missouri is that of a deep drab colour containing a greater portion of Mud than the Rochejhone.

LEWIS / *[August 3, 1806]*

... an establishment on this [Yellowstone] river at clarks Fork ... may therefore be looked to as one of the most important establishments of the western fur trade.... This river can be navigated to greater advantage in perogues than any other craft ... nor is there any of those moving sand bars so formidable to the navigation of many parts of the Missouri....

CLARK / *Wednesday 4th August 1806*

... deturmine to proceed on to a more eliagiable Spot on the Missouri below at which place the Musquetors will be less troublesom and Buffalow more plenty.... wrote a note to Capt Lewis informing him of my intentions and tied it to a pole.... The child of Shabono has been so much bitten by the Musquetors that his face is much puffed up & Swelled....

CLARK / *Thursday 5th August 1806*

The Musquetors was so troublesom .... I set out at an early hour intending to proceed to some other Situation.... I landed on a sand bar from the South Point intending to form a Camp at this place and continue untill Capt Lewis should arive....

CLARK / *Wednesday 11th August 1806*

... here I found two men from the illinoies   Jos. Dixon [Dickson] and ... Handcock [Forrest Hancock] those men are on a trapping expedition up the River Rochejhone. they inform me that they left the Illinois in the Summer 1804.... that the Mandans and Menitarrais wer at war with the Ricaras .... the Assinniboins were also at war with the Mandans &c. and had prohibited the N W. traders from comeing to the Missouri to trade.... Those dificulties if true will I fear be a bar to our expectations of having the Mandan Minetarra & Ricara chief to acompany us to the U. States. Tho we shall endeaver to bring about a peace....

CLARK / *Thursday 12th August 1806*

... at Meridian Capt Lewis hove into Sight with the party which went by the way of the Missouri as well as that which accompanied him from Travellers Rest on Clarks river; I was alarmed on the landing of the Canoes to be informed that Capt Lewis was wounded by an accident. I found him lying in the Perogue, he informed me that his wound was slight and would be well in 20 or 30 days   this information relieved me very much. I examined the wound and found it very bad flesh wound   the ball had passed through the fleshey part of his left thy below the hip bone and cut the cheek of the right buttock for 3 inches in length and the debth of the ball.... After Capt Lewis and myself parted at Trevellers rest, he ... bore his course to the N E untill he Struck Meadicin [Sun] river near where that river Enters the rocky Mts and proceeded down Medicine river to the Missouri .... this rout is a very good one tho ... the best rout would be from the falls of the Missouri by fort mountain ... to the gap through which the great road passes the dividing Mountain .... The total distance from the falls of the Missouri to Clarks river is only 150 miles of a tolerable road ...

CLARK / *Wednesday (Friday) 13th August 1806*

The last night was very cold with a Stiff breeze from the N.W. all hands were on board and we Set out at sunrize and proceeded on very well with a Stiff breeze astern the greater part of the day. passed the enterance of the Little Missouri river at 8 A.M. and arived at the Enterance of Myry [Big Muddy] river at sun set and encamped on the N E Side haveing came by the assistance of the wind, the current and our oars 86 miles. . . . the Misquetors are not so troublesom this evening as they have been. . . .

ORDWAY / *Thursday 14th August 1806.*

. . . about 9 A.M. we arived at our old neighbours the Grousevauntars and Mandans. we Saluted them by firing our Swivvel and blunderbusses a nomber of times   they . . . were verry glad to see us . . . . Capt Lewis fainted as Capt Clark was dressing his wound, but Soon came too again.

CLARK / *Thursday (Saturday) 14th August 1806*

Set out at Sunrise and proceeded on. when we were opposit the Minetares Grand Village we Saw a number of the Nativs viewing of [us]   we derected the Blunderbuses fired Several times, Soon after we Came too at a Croud of the nativs on the bank opposit the Village of the Shoe Indians or Mah-har-has at which place I saw the principal Chief of the Little Village of the Menitarre & the principal Chief of the Mah-har-has. those people were extreamly pleased to See us. the chief of the little Village of the Menetarras cried Most imoderately, I enquired the cause and was informed it was for the loss of his Son who had been killed latterly by the Blackfoot Indians. . . . I walked up to the Black Cats village & eate some simnins [pumpkins] with him, and Smoked a pipe   this Village I discovered had been rebuilt sin[c]e I left it and much smaller than it was; enquiring into the cause was informed that a quarrel had taken place and (a number of) Lodges had removed to the opposd Side. . . .

ORDWAY / *Friday 15th August 1806.*

a clear pleasant morning. Some of the party went at dressing themselves deer Skins &C. the natives brought us corn and beans &C. they brought us a breakfast of boild siniblins and beans &C. the 2 village of Mandans gave us Considerable of corn and more than we would take away. . . . the chiefs of the 1st village wished us to Stay 1 or 2 days longer with them   we gave the Swivvel to the Big Belleys or Grousevauntars.

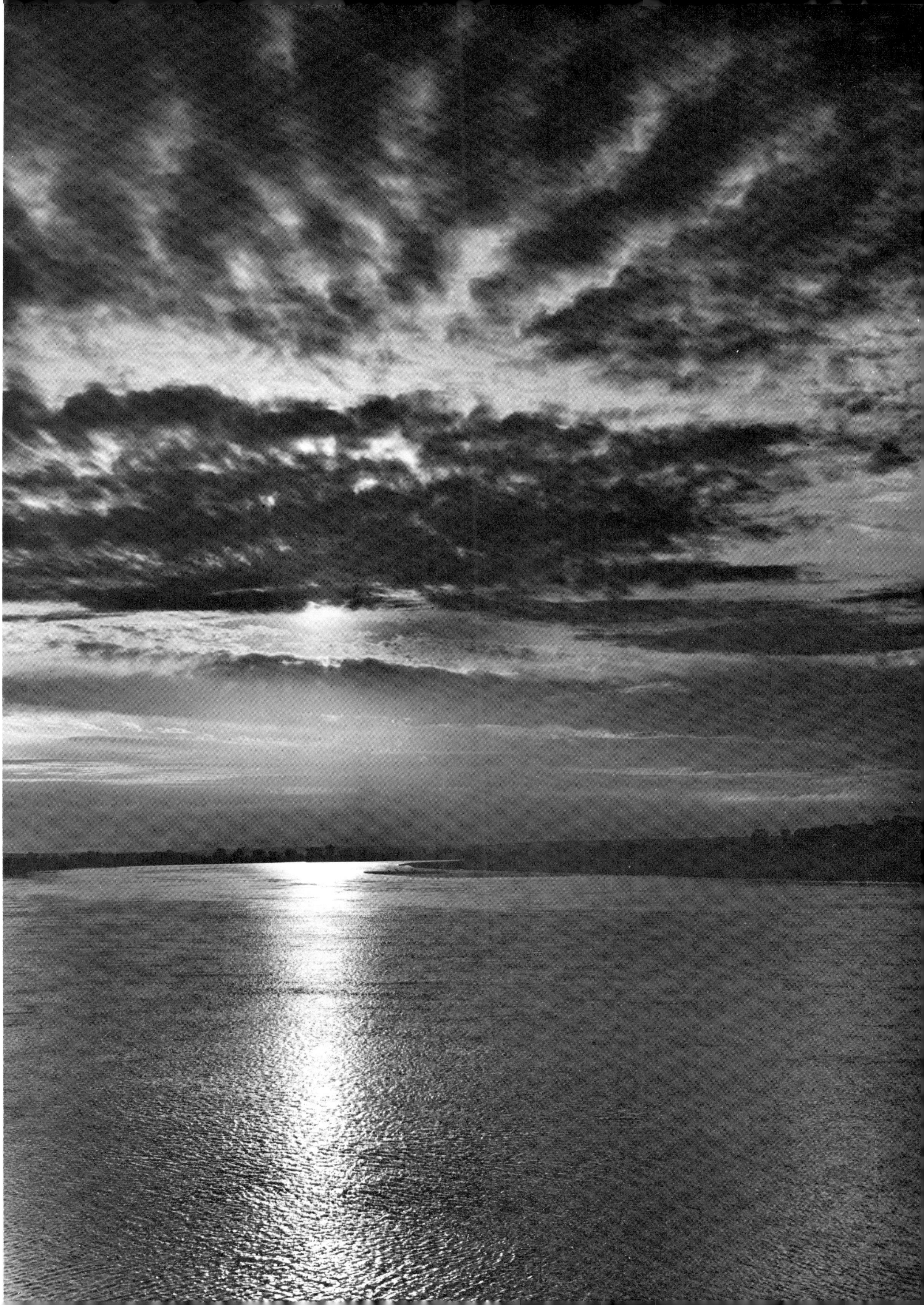

CLARK / Thursday August 15th 1806
Mandans Vilg.

after assembling the Chiefs and Smokeing one pipe . . . . we then envited them to visit their great father the president of the U. States . . . .

The Black Cat Chief of the Mandans Village . . . . informed me that as the Scioux were very troublesom and the road to his great father dangerous none of this village would go down with us. I told the Cheifs and wariers of the village who were then present that we were anxious that some of the village Should go and See their great father and hear his good words & receve his bountifull gifts &c. . . . they hung their heads and said nothing for some time when the Cheif spoke and Said that they were afraid to Send any one for fear of their being killed by the Sieux. . . . Colter one of our men expressed a desire to join Some trappers [the two Illinois Men we met, & who now came down to us] who offered to become shearers with [him] and furnish traps &c. the offer [was] a very advantagious one, to him . . . we agreed to allow him the privilage provided no one of the party would ask or expect a Similar permission to which they all agreeed that they wished Colter every suckcess . . . .

CLARK / Friday 16th August 1806

. . . as our swivel could no longer be Serveceable to us as it could not be fireed on board the largest Perogue, we concluded to make a present of it to the Great Chief of the Menitaras (the One Eye) [Le Borgne] with a view to ingratiate him more Strongly in our favour . . . . the big white Chief would go if we would take his wife & son . . . .

CLARK / Saturday 17th of August 1806

a cool morning gave some powder & Ball to Big White Chief [Shahaka]   Settled with Touisant Chabono for his services as an enterpreter the price of a horse and Lodge purchased of him for public Service in all amounting to 500$ 33⅓ cents. . . .

we were visited by all the principal Chiefs of the Menetarras to take their leave of us   at 2 oClock we left our encampment after takeing leave of Colter . . . . we also took our leave of T. Chabono, his Snake Indian wife and their child [son] who had accompanied us on our rout to the pacific ocean . . . . I offered to take his little son a butifull promising child who is 19 months old to which they both himself & wife wer willing provided the child had been weened. They observed that in one year the boy would be sufficiently old to leave his mother & he would then take him to me if I would be so freindly as to raise the child for him in such a manner as I thought proper, to which I agreeed &c. we droped down to the Big White Cheifs Mandan village . . . . he informed me that he was ready and we were accompd to the Canoes by all the village   Maney of them Cried out aloud. . . . the Cheifs informed that when we first came to their Country they did not beleive all we Said   we then told them. but they were now convinced that everything we had told them was true, that they should keep in memory every thing which we had said to them, and Strictly attend to our advice, that their young men Should Stay at home and Should no[t] go again to war against any nation . . . .

. . . we then saluted them with a gun and set out and proceeded on to Fort Mandan where I landed and went to view the old works   the houses except one in the rear bastion was burnt by accident, some pickets were standing in front next to the river. we proceeded on to the old Ricara village . . . .

CLARK / Monday 18th August 1806

. . . The winds blew hard from the S.E. all day which retarded our progress very much   after the fires were made I set my self down with the bigwhite man Chiefe and made a number of enquiries . . . . he told me his nation first came out of the ground where they had a great village. a grape vine grew down through the Earth to their village and they Saw light Some of their people assended by the grape vine upon the earth, and saw Buffalow and every kind of animal also Grapes plumbs &c. they gathered some grapes & took down the vine to the village, and they tasted and found them good, and deturmined to go up and live upon earth, and great numbers climbed the vine and got upon earth men womin and children. at length a large big bellied woman in climbing broke the vine and fell and all that were left in the village below has remained there ever since (The Mandans beleive when they die that they return to this village)   Those who were left on earth made a village on the river below and were very noumerous &c. . . .

Clark / Tuesday 19th of August 1806

Some rain last night and this morning the wind rose and blew with great Violence untill 4 P.M. . . . Jessomme [René Jessaume] the Interpreter let me have a piece of a lodge and the Squars pitched or Stretched it over Some Sticks, under this piece of leather I Slept dry, it is the only covering which I have had Sufficient to keep off the rain Since I left the Columbia. . . . we decended only 10 Miles to day . . . .

Clark / Wednesday 20th of August 1806

. . . I observe a great alteration in the Current course and appearance of this p.t of the Missouri. in places where there was Sand bars in the fall 1804 at this time the main current passes, and where the current then passed is now a Sand bar. . . .

Clark / Thursday 21t.e August 1806

. . . Met three frenchmen Comeing up . . . inform.d us that 700 Seeoux had passed the Ricaras on their way to war with the Mandans and Menitarras . . . the . . . Ricara Chief who went to the United States last Spring was a year, died on his return . . . .

Clark / Friday 22nd August 1806

. . . as I was about to leave the cheifs [of the Chyennes] lodge he requested me to Send Some traders to them, that their country was full of beaver and they would then be encouraged to kill beaver . . . . I promised the Nation that I would inform their Great father the President of the U States, and he would have them Supplied with goods, and mentioned in what manner they would be Supplied &c. &c.

I am happy to have it in my power to Say that my worthy friend Cap.t Lewis is recovering fast, he walked a little to day for the first time . . . .

I have before mentioned that the Mandans Maharhas Menetarras & Ricarras, keep their horses in the Lodge with themselves at night.

August 17, 1806

CLARK / *Tuesday 26th of August 1806*

... passed the place the Tetons [Sioux] were encamped at the time they attempted to Stop us in Sept 1804.... suspect that the Tetons are not on the Missouri at the big bend as we were informed by the Ricaras, but up on the Teton river.... we made 60 miles to day ....

CLARK / *Friday 29th August 1806*

... I had a view of a greater number of buffalow than I had ever seen before at one time. I must have seen near 20,000 of those animals feeding on this plain. I have observed that in the country between the nations which are at war with each other the greatest numbers of wild animals are to be found....

August 26, 1806

CLARK / *Saturday 30th of August 1806.*

... as we were about to land at the place appointed to wait for the 2 fields [Joseph and Reuben] and Shannon, I saw Several men on horseback which with the help of a spie glass I found to be Indians on the high hills to the N.E. ... I then derected the man who could speak a fiew words of Seioux to inquire what nation or tribe they belong to   they informed me that they were Tetons and their chief was Tar-tack-kah-sab-bar or the black buffalow.... I told this man to inform his nation that we had not forgot their treatment to us as we passed up this river &c that they had treated all the white people who had visited them very badly; robed them of their goods, and had wounded one man whom I had Seen. we viewed them as bad people and no more traders would be Suffered to come to them, and whenever the white people wished to visit the nations above they would come sufficiently Strong to whip any vilenous party who dare to oppose them and words to the same purpote. I also told them that I was informed that a part of all their bands were going to war against the Mandans &c, and that they would be well whiped as the Mandans & Minitarres &[c] had a plenty of Guns Powder and ball, and we had given them a cannon to defend themselves.... 7 of them halted on the top of the hill and blackguarded us, told us to come across and they would kill us all &c of which we took no notice. we all this time were extreamly anxious for the arival of the 2 fields & Shannon whome we had left behind,

September 4, 1806

and were some what cons.^d as to their Safty. to our great joy those men hove in Sight at 6 P. M. . . . we then Set out . . .

CLARK / Wednesday 3rd September 1806

. . . at half past 4 P.M we Spied two boats & Several men . . . I landed and was met by a Mr James Airs from Mackanaw by way of Prarie Dechien and St Louis. . . . our first enquirey was after the President of our country and then our friends and the State of the politicks of our country &c. and the State [of] Indian affairs to all of which enquireys Mr Aires gave us as Satisfactory information as he had it in his power to have collected in the Illinois which was not a great deel. . . . he also informed us that Genl Wilkinson was the governor of the Louisiana [Territory] and at St Louis. . . . the Spaniards had taken one of the U. States frigates in the Mediteranean, Two British Ships of the line had fired on an American Ship in the port of New York, and killed the Capt.^s brother. 2 Indians had been hung in St Louis for murder . . . . and that Mr Burr & Genl Hambleton fought a Duel, the latter was killed &c. &c. I am happy to find that my worthy friend Capt L's is so well as to walk about with ease to himself &c. we made 60 Miles to day

the river much crowded with Sand bars, which are very differently Situated from what they were when we went up.

CLARK / Thursday 4th September 1806.

. . . we came too at Floyds Bluff below the Enterance of Floyds river and assended the hill, with Capt Lewis and Several men, found the grave had been opened by the nativs and left half covered. we had this grave completely filled up, and returned to the canoes and proceeded on to the Sand bar on which we encamped from the 12th to the 20th of August 1804 near the Mahar Village . . . . I observed near Serg.^t Floyds Grave a number of flurishing black walnut trees, these are the first which I have seen decending the river. . . .

CLARK / Saturday 6th September 1806.

The Musquetors excessively troublesom . . . met a tradeing boat of M.^r Og. Choteaux [René Auguste Chouteau] of St Louis bound to the River Jacque to trade with the Yanktons, this boat was in care of a Mr. Henry Delorn . . . . we purchased a gallon of whiskey of this man (promised to pay Choteau who would not receive any pay) *and gave to each man of the party a dram which is the first spiritious liccuor which had been tasted by any of them since the 4 of July 1805.* several of the party exchanged leather for linen Shirts and beaver for corse hats. Those men could inform us nothing more than that all the troops had mov.^d from the Illinois and that Genl Wilkinson was prepareing to leave St Louis. We advised this trader to treat the Tetons with as much contempt as possible and stated to him where he would be benefited by such treatment &c . . . .

CLARK / *Sunday 7th September 1806*

... we proceeded on with a Stiff Breeze ahead (note the evaperation on this portion of the Missouri has been noticed as we assended this river .... I am obliged to replinish my ink Stand every day with fresh ink at least 9/10 of which must evaperate.... we came 44 miles to day only.

CLARK / *Munday 8th September 1806*

... The Missouri at this place does not appear to contain more water than it did 1000 Miles above this, the evaperation must be emence; in the last 1000 miles this river receives the water [of] 20 rivers and maney Creeks Several of the Rivers large and the Size of this river or the quantity of water does not appear to increas any.

CLARK / *Tuesday 9th September 1806*

Set out early at 8 A.M.   passed the enterance of the great river Platt which is at this time low the water nearly clear the current turbelant as usial .... our party appears extreamly anxious to get on, and every day appears [to] produce new anxieties in them to get to their country and friends. My worthy friend Cap Lewis has entirely recovered   his wounds are heeled up and he can walk and even run nearly as well as ever he could, the parts are yet tender &c ....

September 8, 1806

CLARK / *Wednesday 10th September 1806.*
... M.r la frost [Alexander La fass] informed us that
... [Zebulon] Pike and young M.r [James D.]
Wilkinson had Set out on an expedition up the
Arkansaw river .... we find the river in this timbered
country narrow and more moveing Sands and a much
greater quantity of Sawyers or Snags than above.
Great caution and much attention is required to Stear
clear of all those dificuelties in this low State of the
water. we made 65 miles to day.... one of the men
killed a racoon which the indians very much admired.

Clark / *Friday 12th of September 1806*
... at S.t Mich.ls Prarie ... we found M.r Jo. Gravelin
the Ricaras enterpreter whome we had Sent down
with a Ricaras Chief in the Spring of 1805. and old
M. Durion the Sieux enterpreter, we examined the
instructions of those interpreters and found that
Gravelin was ordered to the Ricaras with a Speach
from the president of the U. States to that nation and
some presents which had been given the Ricara Chief
who had visited the U. States and unfortunately died
at the City of Washington, he was instructed to teach
the Ricaras agriculture & make every enquirey after
Cap.t Lewis my self and the party. M.r Durion was
enstructed to accompany Gravelin and through his
influence pass him with his presents & [c.] by the
tetons bands of Sieux, and to provale on Some of the
Principal chiefs of those bands not exceeding six to
Visit the Seat of the Government next Spring. he was
also enstructed to make every enquirey after us....

CLARK / *Wednesday 17th September 1806*
we Set out as usial early .... at 11 A. M. we met a
Captain M.cClellin late a Cap.t of Artil.y of the U States
Army assending in a large boat. this gentleman an
acquaintance of my friend Cap.t Lewis was Somewhat
astonished to see us return and appeared rejoiced to
meet us.... we were makeing enquires and
exchangeing answers &c. untill near mid night. this
Gentleman informed us that we had been long Since
given out [up] by the people of the U S Generaly
and almost forgotton, the President of the U. States
had yet hopes of us ... he gave us Some Buisquit,
Chocolate Sugar & whiskey, for which our party were
in want and for which we made a return of a barrel of
corn & much obliged to him. Cap.t M.cClellin
informed us that he was on reather a speculative
expedition to the confines of New Spain, with the
view to entroduce a trade with those people....

September 10, 1806

CLARK / *Friday 19th of Sept. 1806.*

Set out this morning a little after day & proceeded on very well   the men plyd their oares & we decended with great velocity, only came too once for the purpose of gathering pappows, our anxiety as also the wish of the party to proceed on as expeditiously as possible to the Illinois enduce us to continue on without halting to hunt. we calculate on ariveing at the first Settlements on tomorrow evening which is 140 miles . . . . three of the party have their eyes inflamed and Sweled in Such a manner as to render them extreamly painfull, particularly when exposed to the light, the eye ball is much inflaimed . . . . I am willing to believe it may be owing to the reflection of the sun on the water.

CLARK / *Saturday 20th Sept. 1806*

as three of the party was unabled to row from the State of their eyes we found it necessary to leave one of our crafts and divide the men into the other Canoes . . . . the party being extreemly anxious to get down ply their ores very well, we saw some cows on the bank which was a joyfull Sight to the party and caused a Shout to be raised for joy . . . .

September 20, 1806

September 21, 1806

CLARK / Sunday 21st Sepr 1806
... at half after 7 A. M we Set out. passed 12 canoes of Kickapoos assending on a hunting expedition. Saw Several persons also stock of different kind on the bank which reviv'd the party very much. ... at 4 P M we arived in Sight of St Charles, the party rejoiced at the Sight of this hospita[b]l[e] village plyed thear ores with great dexterity and we Soon arived opposit the Town   this day being Sunday we observed a number of Gentlemen and ladies walking on the bank, we saluted the Village by three rounds from our blunderbuts and the Small arms of the party, and landed near the lower part of the town. we were met by great numbers of the inhabitants, we found them excessively polite. we received invitations from Several of those Gentlemen .... the inhabitants of this village appear much delighted at our return and seem to vie with each other in their politeness to us all. we came only 48 miles to day. the banks of the river thinly settled &c. (some Settlements since we went up)

CLARK / *Monday 22nd of Sept. 1806*

... we did not [think] proper to proceed on untill after the rain was over, and continued at the house of M.r Proulx. I took this oppertunity of writeing to my friends in Kentucky &c. at 10 A M. it seased raining and we colected our party and Set out and proceeded on down to the Contonem.t at Coldwater Creek .... M.rs Wilkinson the Lady of the Gov.r & Gen.l we wer sorry to find in delicate health.

we were honored with a Salute of ... Guns and a harty welcom. at this place there is a publick store kept in which I am informed the U.S have 60000$ worth of indian Goods

CLARK / *Tuesday 23rd Sept.r 1806*

we rose early took the Chief [Shahaka] to the publick store & furnished him with Some clothes &c. took an early breckfast with Col.o Hunt and Set out decended to the Mississippi and down that river to S.t Louis at which place we arived about 12 oClock. we Suffered the party to fire off their pieces as a Salute to the Town. we were met by all the village and received a harty welcom from it's inhabitants &.c. ... we payed a friendly visit to M.r August Chotau and some of our old friends this evening. as the post had departed from S.t Louis Cap.t Lewis wrote a note to M.r Hay in Kahoka [Cahokia] to detain the post at that place untill 12 tomorrow which was reather later than his usial time of leaveing it

ORDWAY / *Tuesday 23rd Sep.t 1806.*

... about 12 oClock we arived in Site of S.t Louis fired three Rounds as we approached the Town and landed oppocit the center of the Town .... drew out the canoes then the party all considerable much rejoiced that we have the Expedition Completed and now we look for boarding in Town and wait for our Settlement and then we entend to return to our native homes to See our parents once more as we have been so long from them.—finis.

GASS / *[23 September 1806]*

... anxious to reach St. Louis, where, without any important occurrence, we arrived on the 23.d, and were received with great kindness and marks of friendship by the inhabitants, after an absence of two years, four months, and ten days.

CLARK / *Wednesday 24.th of September 1806*

I sleped but little last night   however we rose early and commenc[e]d wrighting our letters Cap.t Lewis wrote one to the presidend and I wrote Gov.r Harrison & my friends in Kentucky and Sent of[f] George Drewyer with those letters to Kohoka & delivered them to M.r Hays &c. we dined with M.r Chotoux [Chouteau] to day, and after dinner went to a store and purchased some clothes, which we gave to a Tayler and directed to be made. Cap.t Lewis in opening his trunk found all his papers wet, and some seeds spoiled.

CLARK / *Thursday 25.th of Sept.r 1806*

had all of our skins &c. suned and stored away in a storeroom of M.r Caddy Choteau. payed some visits of form, to the gentlemen of S.t Louis. in the evening a dinner & Ball

CLARK / *Friday 25.th [26] of Sept.r 1806*

a fine morning   we commenced wrighting &c.

Perminent Members of the Expedition
    Officers
        Meriwether Lewis
        William Clark
    Sergeants
        Patrick Gass
        John Ordway
        Nathaniel Pryor
    Privates
        William Bratton
        John Collins
        John Colter
        Peter Cruzatte
        Joseph Fields
        Reuben Fields
        Robert Frazier
        George Gibson
        Silas Goodrich
        Hugh Hall
        Thomas P. Howard
        Francis Labiche
        Baptiste Lepage
        Hugh McNeal
        John Potts
        George Shannon
        John Shields
        John B. Thompson
        William Werner
        Joseph Whitehouse
        Alexander Willard
        Richard Windsor
        Peter Wiser
    Interpreters
        George Drouillard
        Toussaint Charbonneau
    Other Members
        Sacajawea and her infant son,
            Baptiste Charbonneau
        York, Negro slave

[*December 2, 1806*]

Gentlemen of the Senate and of the House of Representatives.

The expedition of Messrs. Lewis and Clarke, for exploring the river Missouri, and the best communication from that to the Pacific Ocean, has had all the success which could have been expected. They have traced the Missouri nearly to its source, descended the Columbia to the Pacific ocean, ascertained with accuracy the georgraphy of that interesting communication across our continent, learned the character of the country, of its commerce, and inhabitants; and it is but justice to say that Messrs. Lewis and Clarke, and their brave companions, have by this arduous service deserved well of their country.

[Thomas Jefferson, Annual Message to Congress]

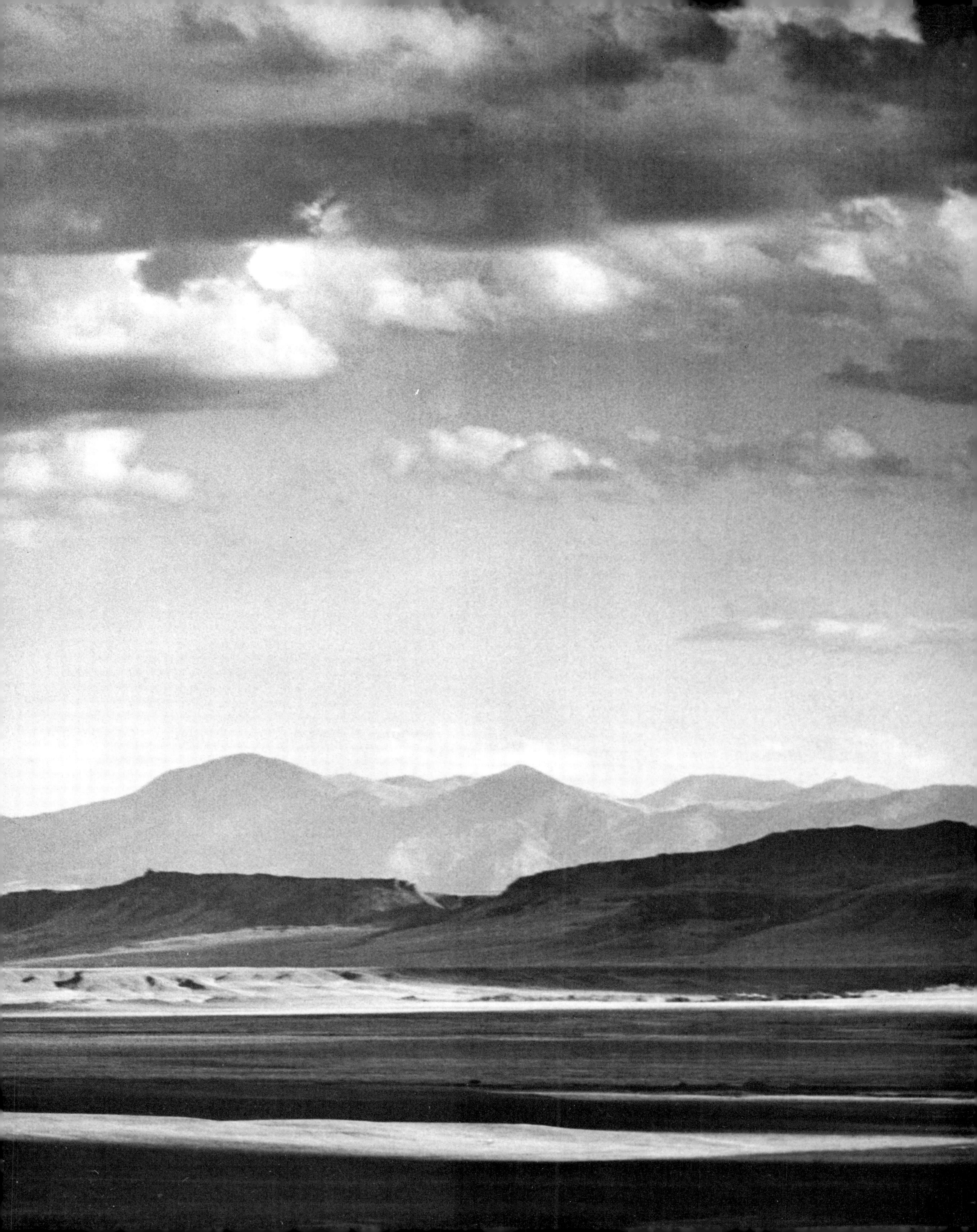

LIST OF ILLUSTRATIONS

	Illustration
Cover	*Lewis and Clark Meeting Indians at Ross' Hole* by Charles M. Russell. Courtesy of the Montana Historical Society

Page	**Journal Entry**	**Illustration**
i		Missouri River, south of Omaha, Nebraska
ii-iii		The Pinnacles, Missouri River Breaks, Montana
iv		Blue and Canada Geese, near the Squaw Creek, Missouri
v		Pack Creek, near Packer Meadow, Montana-Idaho border
vi		Detail, Missouri River Breaks, Montana
vi		Squirrel
vi		Burr Oak, near Columbia, Missouri
vii		Bear Grass, Clearwater National Forest, Idaho
vii		Bighorn Sheep
viii		Meriwether Lewis and William Clark
x		Missouri River Breaks, Montana
xi		Cedar Tree, Bernard DeVoto Memorial Grove, Crooked Fork, Clearwater National Forest, Idaho
xii		Ursula Higgins, Browning, Montana
xii-xiii		The Pinnacles, Missouri River Breaks, Montana
xiv		Missouri River, northeast of Virgelle, Montana
xvi		Missouri River Breaks, Montana

231

Page	Journal Entry	Illustration
xviii		Crown Butte, near Fort Shaw, Montana
xxi		Detail, Clearwater River, Idaho
1		Jefferson Peace Medal, front
5		Flintlock Blunderbuss, from 1901-04 Biennial Report, Montana State Game and Fish Warden
5		Buffalo Skull, found near Cut Bank, Montana
8		Engraving, Falls of the Ohio River
10		Falls of the Ohio River, near Louisville, Kentucky
11		Masthead, *Kentucky Gazette and General Advertiser*
12		Silver Creek, Clarksville, Indiana
13		Confluence, Missouri and Mississippi rivers, Lewis and Clark Memorial Park, Wood River, Illinois
14		Map, Illinois Country
20-1		A Map of Lewis and Clark's Track Across the Western Portion of North America, from United States National Archives
22	May 16, 1804	Buildings, St. Charles, Missouri
24	May 23, 1804	Tavern Cave, near St. Albans, Missouri
25	June 4, 1804	Burr Oak, near Columbia, Missouri
26	June 7, 1804	Pictograph Rocks, Rocheport, Missouri
27	June 7, 1804	Pictograph Rocks, Rocheport, Missouri
28	July 4, 1804	Missouri River, near Atchison, Kansas
29 (top)	July 4, 1804	Blue and Canada Geese, near the Squaw Creek, Missouri
29	July 4, 1804	Independence Creek, near Doniphan, Kansas
30 (top)	July 7, 1804	Prairie du Chien, near St. Joseph, Missouri
30	July 12, 1804	Nemaha River, Rulo, Nebraska
31	July 21, 1804	Missouri River, Plattsmouth Waterfowl Management Area, near Plattsmouth, Nebraska
32-3	July 30, 1804	Council Bluff, south of Blair, Nebraska
33	July 31, 1804	Beaver
34	August 11, 1804	Blackbird Hill, Omaha Indian Reservation, Nebraska
35	August 20, 1804	Floyd's Bluff, near Sergeant Floyd Monument, Sioux City, Iowa
35	September 7, 1804	Prairie Dog
36	September 14, 1804	Antelope
36	September 20, 1804	Reuben Creek, Big Bend of the Missouri, near De Grey, South Dakota
37	September 23, 1804	Campsite, south of De Grey, South Dakota

Page	Journal Entry	Illustration
38 (top)	October 18, 1804	Cannonball River, near Fort Rice, North Dakota
38	October 19, 1804	Knolls, near Fort Clark, North Dakota
39	October 19, 1804	Buffalo
40	October 21, 1804	Heart River, at Mandan, North Dakota
41	October 26, 1804	Ice on Knife River, Sakakawea Village Site, south of Garrison Dam, North Dakota
42	October 27, 1804	Mandan Village, Slant Indian Village Site, Mandan, North Dakota
45 (top)	November 18, 1804	River Ice and Tree Trunk, near Sanger, North Dakota
45	January 7, 1805	Deer
46 (top)	January 13, 1805	Buffalo
46	February 5, 1805	Battle Ax, found near Stanton, North Dakota
49	April 8, 1805	High Bluff, Black Cat's Camp, Fort Berthold Reservation, North Dakota
50	April 25, 1805	Glass's Bluff, Missouri River, at the Montana-North Dakota border
50-1	April 26, 1805	Moonlight, Confluence, Yellowstone and Missouri rivers, near Williston, North Dakota
51	May 9, 1805	Buffalo Herd
52 (left)	May 11, 1805	Grizzly Bear
52	May 20, 1805	Horned Owl
52 (bottom)	May 17, 1805	Snake
53	May 24, 1805	Ship Rock, on Missouri River Breaks, Montana
54 (top)	May 26, 1805	Rocky Mountains, from east of Judith River mouth, Montana
54	May 26, 1805	Point of Rocks, Missouri River Breaks, Montana
55	May 26, 1805	Hole-in-the-Wall, Missouri River Breaks
56	May 27, 1805	Sandstone Erosion, Missouri River Breaks
57	May 27, 1805	Missouri River Breaks
58-9	May 27, 1805	Black Butte, from Missouri River Breaks
59	May 27, 1805	Hole-in-the-Wall, Missouri River Breaks
60	May 28, 1805	Dog Creek, Missouri River Breaks
61	May 29, 1805	Buffalo
62	May 29, 1805	Black Butte, from the Judith River, Montana
63 (top)	May 29, 1805	Camp at Judith River, Montana
63	May 29, 1805	Camp at Arrow Creek, Montana
64-5	May 30, 1805	Camp at Point of Rocks, Missouri River Breaks
65	May 30, 1805	Column, Missouri River Breaks

Page	Journal Entry	Illustration
66	May 30, 1805	Missouri River Breaks
67 (top)	May 31, 1805	The Castle, Missouri River Breaks
67	May 31, 1805	Cathedral Carvings, Missouri River Breaks
68 (top)	May 31, 1805	Missouri River Breaks
68	May 31, 1805	The Pinnacles, Missouri River Breaks
69	May 31, 1805	Citadel Rock, Missouri River Breaks
70	May 31, 1805	Swallows' Nests, Missouri River Breaks
71 (top)	May 31, 1805	Cliffs at Eagle Creek, Missouri River Breaks
70-1	May 31, 1805	Missouri River Breaks, near Eagle Creek
72	May 31, 1805	Camp at Eagle Creek, Missouri River Breaks
72	May 31, 1805	Camp at Eagle Creek, Missouri River Breaks
73	June 8, 1805	Marias River, Loma, Montana
74-5	June 14, 1805	Great Falls of the Missouri, Great Falls, Montana
75	June 14, 1805	Great Falls of the Missouri, Great Falls, Montana
76		Clark's Map of the Portage
77	June 29, 1805	Giant Springs, Great Falls, Montana
78-9	June 29, 1805	Black Clouds, Continental Divide, near Fort Shaw, Montana
80	July 2, 1805	White Bear Island, above Great Falls, Montana
81	July 7, 1805	Elk
82 (top)	July 13, 1805	Canoe Camp, near Great Falls, Montana
82	July 15, 1805	Square Butte [Fort Mountain], near Fort Shaw, Montana
83	July 16, 1805	Approaching the Rockies, southwest of Great Falls, Montana
84	July 19, 1805	Gate of the Mountains, Helena National Forest, near Helena, Montana
85 (top)	July 19, 1805	Gate of the Mountains, near Camp Meriwether, Helena National Forest, Helena, Montana
85	July 19, 1805	South Channel, Gate of the Mountains, Helena National Forest, near Helena, Montana
86 (left)	July 24, 1805	Prickly Pear
86	July 25, 1805	Geese
86-7	July 27, 1805	Missouri River, Southeast Fork, at Three Forks, Montana
87	July 27, 1805	Limestone Cliff, Three Forks Recreation Site, Three Forks, Montana
88	August 8, 1805	Point of Rocks, north of Dillon, Montana
89 (top)	August 10, 1805	Rattlesnake Cliffs, south of Dillon, Montana
89	August 10, 1805	Beaverhead Rock, south of Dillon, Montana

Page	Journal Entry	Illustration
90	August 12, 1805	Fountainhead of the Missouri, Lemhi Pass, Beaverhead National Forest, Continental Divide, Montana
91	August 12, 1805	Idaho from Lemhi Pass, Continental Divide, Montana
92	August 13, 1805	Three Indians, south of Billings, Montana
94	August 20, 1805	Lemhi River, north of Tendoy, Idaho
94-5	August 21, 1805	Elk
96	August 23, 1805	Lost Rapids, Salmon River, Salmon National Forest, near North Fork, Idaho
97 (top)	August 23, 1805	Salmon River, Salmon National Forest, near North Fork, Idaho
97	August 23, 1805	Impenetrable Cliffs, Salmon National Forest, North Fork, Idaho
98	August 23, 1805	Mule Deer
99	August 23, 1805	Salmon River Mountains, Salmon National Forest, Idaho
100	August 29, 1805	Approaching the Bitterroot Range, near Gibbonsville, Idaho
101	August 31, 1805	The Towers, between Gibbonsville and North Fork, Idaho
101	August 31, 1805	The Towers, between Gibbonsville and North Fork, Idaho
102	September 3, 1805	Lost Trail Pass, Salmon and Bitterroot National Forests, Continental Divide, Idaho
103 (top)	September 4, 1805	Tushapaw [Flathead] Man, Arlee, Montana
103 (middle)	September 4, 1805	Tushapaw [Flathead] Woman, Arlee, Montana
103	September 7, 1805	Trapper Peak, Bitterroot National Forest, south of Darby, Montana
104 (top)	September 9, 1805	Sapphire Mountains, Bitterroot National Forest, near Florence, Montana
104	September 9, 1805	Travelers Rest, Lolo, Montana
105 (top)	September 10, 1805	Chief Charles [Flathead], Arlee, Montana
105	September 10, 1805	Louis Calooyia [Flathead], Arlee, Montana
106-7	September 11, 1805	Lolo Peak, Lolo National Forest, near Lolo, Montana
107	September 12, 1805	Tushapaw [Flathead] Sweathouse
108	September 13, 1805	Packer Meadow Camp, Lolo Pass, Continental Divide, Montana
109	September 14, 1805	Island in Kooskooskee [Clearwater] River, Powell Ranger Station, Clearwater National Forest, Idaho
110 (left)	September 15, 1805	Mountain Spring, Clearwater National Forest
110	September 15, 1805	Approaching Lolo Indian Trail, Clearwater National Forest

Page	Journal Entry	Illustration
111 (top)	September 15, 1805	Granitic Rock, Indian Graves Lookout, Clearwater National Forest
111	September 15, 1805	Lolo Indian Trail, Clearwater National Forest
112	September 16, 1805	Cloudy Morning, Clearwater National Forest
113 (top)	September 17, 1805	Balsams, Clearwater National Forest
113	September 18, 1805	Sundown on Lolo Indian Trail, near Jerry Johnson Lookout, Clearwater National Forest
114-15	September 19, 1805	Weippe Prairie, from Wood Rat Lookout, Clearwater National Forest
116	September 20, 1805	Josiah Redwolf [Nez Percé], Spalding, Idaho
117	September 21, 1805	Camp of Twisted Hair, on Clearwater River, Orofino, Idaho
116-17	September 20, 1805	Camas Bulbs, Clearwater National Forest
118	September 26, 1805	Canoe Camp, on Clearwater River, near Orofino, Idaho
119	October 5, 1805	Branding Iron
120-1	October 6, 1805	Kooskooskee [Clearwater] River, near Orofino, Idaho
120	October 11, 1805	Tushapaw [Flathead] Bathhouse
121	October 12, 1805	Snake River, south of Kahlotus, Washington
122	October 13, 1805	Palouse River, east of Kahlotus, Washington
123	October 14, 1805	Hull of a Ship Rock, south of Kahlotus, Washington
124	October 15, 1805	Snake River, south of Kahlotus, Washington [Note: This scene along the Snake River has importance for its relationship to Lewis and Clark's journey. However, a huge earthen dam fills the site and there was no possible way to recapture their view except to technically "approximate" the scene. This I have done, with no intention of deceit.]
125	October 16, 1805	Confluence, Snake and Columbia rivers, near Sacajawea State Park, Pasco, Washington
126	October 18, 1805	Columbia River, Lake Wallula, Washington
127	October 18, 1805	Walls at the Gap, near Wallula, Washington
128-9	October 18, 1805	Captains' Rock, Columbia River, near Wallula, Washington
130 (top)	October 19, 1805	Hat Rock, Hat Rock State Park, near Umatilla, Oregon

Page	Journal Entry	Illustration
130	October 21, 1805	Columbia River, Celilo, Oregon
131	October 21, 1805	Miller's Island, near Celilo, Oregon
132	October 21, 1805	Mount Hood, Mount Hood National Forest, from The Dalles, Oregon
132-3	October 22, 1805	The Dalles, Columbia River, east of Celilo Falls, in Oregon
134	October 22, 1805	Great Falls of the Columbia, The Dalles, Oregon
134-5	October 24, 1805	Basin near the Narrows, The Dalles, Oregon
136 (top)	October 25, 1805	The Dalles, Columbia River, Washington
136	October 25, 1805	Columbia River Wall, Columbia Gorge, Cascade Range, Washington
137	October 31, 1805	Beacon Rock, Che-che-op-tin, Columbia River, Beacon Rock State Park, Skamania, Washington
138-9	November 3, 1805	Mount Hood, Mount Hood National Forest, from Hood River, Oregon
140-1	November 7, 1805	Grays Bay, Columbia River, near Rosburg, Washington
142	November 13, 1805	Point Ellice, near Fort Columbia National Park, Chinook, Washington
143	November 14, 1805	Sand Beach, near Fort Columbia State Park, Chinook, Washington
144	November 15, 1805	Objective of the Expedition, Lewis and Clark State Park, near Chinook, Washington
145	November 16, 1805	Trees, near Cannon Beach, Oregon
146	November 18, 1805	Coastline, near Cannon Beach, Oregon
147	November 18, 1805	Rain Forest, near Chinook, Washington
149	November 19, 1805	Beach, near Long Beach, Washington
150	November 21, 1805	Salmon
151	November 23, 1805	Coastal Detail
152	November 30, 1805	Pacific Ocean, near Fort Canby State Park, Illwaco, Washington
153	December 1, 1805	Pacific Ocean, near Chinook, Washington
154 (top)	December 10, 1805	Ecola [Clark's] Beach, near Ecola State Park, Cannon Beach, Oregon
154	December 30, 1805	Fort Clatsop, Fort Clatsop National Memorial, near Astoria, Oregon
156	January 1, 1806	Fort Clatsop Detail
157	January 5, 1806	Salt Works, at Seaside, Oregon
159	January 8, 1806	Ecola State Park, Cannon Beach, Oregon
160	February 6, 1806	Balsam Fir
161	February 25, 1806	Raccoon Tracks
162-3	March 15, 1806	Clam Shells, Crow Butte Recreation Site, Washington
165	April 1, 1806	Sandy [Quicksand] River, near Gresham, Oregon

Page	Journal Entry	Illustration
165	April 5, 1806	Tree Dogwood, St. Helens, Oregon
166 (top)	April 9, 1806	Multnomah Falls, Multnomah Falls Lodge, Mount Hood National Forest, Columbia River, Oregon
166	April 9, 1806	Beacon Rock, west of Bonneville Dam, at the Oregon-Washington border
167	April 9, 1806	Horse Tail Falls, west of Bonneville Dam, Mount Hood National Forest, Columbia River, Oregon
168	April 18, 1806	Big Basin, Horse Thief Lake, near The Dalles, Oregon
169	April 20, 1806	East of Miller's Island, near Celilo, Oregon
170	April 23, 1806	Miller's Island, Celilo, Oregon
172	April 25, 1806	Willow Camp, Crow Butte Recreation Site, Washington
173	May 7, 1806	Spurs of the Rockies, Clearwater Range, near Nezperce, Idaho
174	May 10, 1806	Clouds in Valley, near Nez Percé Indian Reservation, Nezperce, Idaho
175	May 12, 1806	Appaloosa Horse, Spalding, Idaho
176-7	May 17, 1806	Bear Grass, Lolo Indian Trail, near Powell Ranger Station, Idaho
177	May 17, 1806	Cone Flowers, Lewis and Clark Trail, Clearwater National Forest, Idaho
178	May 27, 1806	Squirrel
178	May 28, 1806	Bighorn Sheep
179	May 29, 1806	Clark's Nutcracker
179		Flower, Lochsa River, Idaho
180-1	June 10, 1806	Clearwater Range, Clearwater National Forest, Idaho
182	June 19, 1806	Morrell Mushroom
182	June 19, 1806	Mules
182-3	June 20, 1806	View from Castle Butte, Clearwater National Forest, Idaho
183 (left)	June 21, 1806	Lolo Indian Trail, Clearwater National Forest, Idaho
183	June 25, 1806	Burned Fir, near Jerry Johnson Lookout, Clearwater National Forest
184 (top)	June 27, 1806	Bear Grass, Clearwater National Forest, Idaho
184	June 27, 1806	Indian Post Office, Clearwater National Forest, Idaho
185	June 28, 1806	Mountain Slopes, Clearwater National Forest, Idaho
186 (left)	June 29, 1806	Lolo Hot Springs, west of Lolo, Montana
186-7	July 1, 1806	Travelers Rest, mouth of Lolo Creek, Lolo, Montana
187	July 2, 1806	Flower

Page	Journal Entry	Illustration
188	July 4, 1806	Flathead Indian
188-9	July 4, 1806	Big Blackfoot River, near Bonner, Montana
190	July 5, 1806	Knobby Hills, Ovando, Montana
191 (top)	July 6, 1806	Blackfoot Prairie, Ovando, Montana
191	July 7, 1806	Moose, at Alice Creek, Helena National Forest, Lincoln, Montana
192 (top)	July 7, 1806	Lewis and Clark Pass, Helena National Forest, Continental Divide, Helena, Montana
192	July 7, 1806	Square Butte [Fort Mountain], near Fort Shaw, Montana
193	July 7, 1806	Montana Rockies, Continental Divide
194 (left)	July 11, 1806	Buffalo Head
194	July 15, 1806	Bear
195	July 17, 1806	Great Falls of the Missouri, Great Falls, Montana
196-7	July 22, 1806	Indian Encounter 1, at Two Medicine River, south of Browning, Montana
197	July 26, 1806	Indian Encounter 2, near Two Medicine Buttes, southwest of Cut Bank, Montana
198 (top)	July 26, 1806	Indian Encounter 3, on Two Medicine River, southwest of Cut Bank, Montana
198	July 26, 1806	Indian Encounter 4
199	July 26, 1806	Indian Encounter 5
200	July 27, 1806	Indian Encounter 6
201	July 27, 1806	Louis Plenty Treaty [Blackfoot], Browning, Montana
202-3	July 27, 1806	Indian Encounter 7, on Two Medicine River, southwest of Cut Bank, Montana
206-7	July 3, 1806	Bitterroot Valley, near Lolo, Montana
206	July 10, 1806	Skull and Rattlesnake
208	July 14, 1806	Bozeman Pass, near Bozeman, Montana
209	July 15, 1806	Bozeman Pass, near Bozeman, Montana
210	July 20, 1806	Dugout Canoes, near Park City, Montana
211 (top)	July 24, 1806	Yellowstone River, view west from Pompey's Pillar
211	July 24, 1806	Young's Point, west of Park City, Montana
212-13	July 25, 1806	Pompey's Pillar, northeast of Billings, Montana
213	July 25, 1806	Signature, Pompey's Pillar
214 (top)	August 3, 1806	Confluence, Yellowstone and Missouri rivers, southwest of Williston, North Dakota
214	August 5, 1806	Missouri River, near Williston, North Dakota
217	August 14, 1806	Sunrise on the Missouri River, North Dakota

Page	Journal Entry	Illustration
219	August 17, 1806	Grand [Mandan] Village, Hidatsa Village Site, north of Bismarck, North Dakota
220	August 26, 1806	Campsite with Teton Sioux, Pierre, South Dakota
221	September 4, 1806	Floyd's Bluff, near Floyd's Monument, Sioux City, Iowa
222	September 8, 1806	Missouri River, near Omaha, Nebraska
223	September 10, 1806	Raccoon Tracks
224	September 20, 1806	Missouri River, La Charrette, Missouri
225	September 21, 1806	Stairway, St. Charles, Missouri
227		Jefferson Peace Medal, back
228-9		Approaching the Rockies, southwest of Great Falls, Montana

A NOTE ON THE PHOTOGRAPHY

My equipment was simple. I employed a Linhof Technika, especially designed for serious photography, using both 4x5 and 5x7-inch film. To be sure, it was slower, since it had to be secured on a tripod, but it had its advantages. I used an assortment of interchangeable lenses for better perspective and quality of illustration. The large film size was a requirement because, working slowly, I needed it for both sharpness and depth. The larger film gave me better print quality.

Eastman Royal Pan film (ASA 400) provided a moderate grain slightly on the soft side. Standard film and paper developer gave me consistent quality from the beginning. Printing was done on Du Pont Varilour, a variable contrast paper. Proof prints which were studied for final objectives and tonal corrections were printed on 11x14-inch paper, and final reproductions were still-made on 11x14 paper and mounted on 14x17 white mounts.

SOURCES

By far the single most important source for *American Odyssey* is the eight-volume *Original Journals of the Lewis and Clark Expedition*. From the invaluable *Original Journals* came the orderly books, the journal of Private Joseph Whitehouse, the journal of Sergeant Charles Floyd, some of the letters, as well as the majority of the entries.

Thwaites, Reuben Gold, ed. *Original Journals of the Lewis and Clark Expedition*. 8 vols. Dodd, Mead & Co., New York, 1904-05.

Some of the early entries by both Lewis and Clark come from different sources. These are identified by date below, as are the entries of Sergeants Ordway and Gass, the entries of Jonathan Clark (brother of William), the news stories from the *Kentucky Gazette and General Advertiser,* and passports and various letters dated prior to the departure of the Expedition.

Jackson, Donald, ed. *Letters of the Lewis and Clark Expedition with Related Documents, 1783-1854*. University of Illinois Press, Urbana, 1962.

    Letters:
        Jefferson to Lewis   February 23, 1801
        Lewis to Jefferson   March 10, 1801
        Martínez de Yrujo to Cevallos   December 2, 1802
    Lewis's British Passport   February 28, 1803
    Lewis's French Passport   March 1, 1803
    Letters:
        Gallatin to Jefferson   April 13, 1803
        Lacépède to Jefferson   May 13, 1803
        Lewis to Lucy Marks   July 2, 1803
        Clark to Jefferson   July 24, 1803
        Delassus to Salcedo and Casa Calvo   December 9, 1803
        Lewis to Jefferson   December 19, 1803
        Lewis to Jefferson   December 28, 1803
        Clark to Croghan   January 15, 1804
        Lewis to Clark   May 6, 1804

Osgood, Ernest Staples, ed. *Field Notes of Captain William Clark, 1803-1805*. Yale University Press, New Haven and London, 1964.

    Clark's Entries of:

December 13, 1803	March 26, 1804
December 16, 1803	March 28, 1804
December 23, 1803	April 7, 1804
December 25, 1803	April 19, 1804
January 2, 1804	April 28, 1804
January 30, 1804	May 7, 1804
February 5, 1804	

Quaife, Milo M., ed. *Journals of Captain Meriwether Lewis and Sergeant John Ordway*. The State Historical Society of Wisconsin, Madison, 1916 (reprinted in 1965).

Lewis's Entries of:
 August 30, 1803 November 11, 1803
 September 7, 1803 November 16, 1803
 September 15, 1803 December 12, 1803
 September 18, 1803

Ordway's Entries of:
 July 6, 1804 August 24, 1805
 September 14, 1804 August 27, 1805
 October 27, 1804 August 30, 1805
 November 5, 1804 September 5, 1805
 November 22, 1804 September 23, 1805
 January 1, 1805 October 14, 1805
 January 14, 1805 October 18, 1805
 January 30, 1805 October 21, 1805
 February 28, 1805 October 22, 1805
 April 7, 1805 December 25, 1805
 May 18, 1805 December 30, 1805
 June 6, 1805 January 6, 1806
 June 8, 1805 March 23, 1806
 June 17, 1805 April 14, 1806
 June 29, 1805 April 16, 1806
 July 4, 1805 May 3, 1806
 July 9, 1805 May 23, 1806
 July 22, 1805 May 28, 1806
 August 9, 1805 July 3, 1806
 August 18, 1805 August 11, 1806
 August 20, 1805 August 14, 1806
 August 22, 1805 September 23, 1806

Gass, Patrick. *A Journal of the Voyages and Travels of a Corps of Discovery.* London, 1808. o.p.

 Entries of:
  May 14, 1804 March 20, 1806
  October 27, 1804 April 21, 1806
  December 16, 1804 May 2, 1806
  January 29, 1805 May 15, 1806
  April 5, 1805 May 25, 1806
  April 26, 1805 June 2, 1806
  May 27, 1805 June 27, 1806
  September 5, 1805 July 16, 1806
  October 7, 1805 August 7, 1806
  October 23, 1805 September 23, 1806

Kentucky Gazette and General Advertiser. Library of the Filson Club, Louisville, Kentucky.

 News Stories of:
  October 15, 1803
  October 22, 1803
  October 29, 1803

Clark, Jonathan. *Journal.* Library of the Filson Club, Louisville, Kentucky.

 Entries of:
  October 26, 1803
  November 12, 1803

BOOKS FOR FURTHER READING

Burroughs, Raymond D. *The Natural History of the Lewis and Clark Expedition.* Michigan State University Press, East Lansing, 1961.

Coues, Elliott, ed. *History of the Expedition Under the Command of Lewis and Clark.* Dover Publishing Co., New York, 1965 (reprinted from the 1895 edition).

Criswell, Elijah Harry. *Lewis and Clark: Linguistic Pioneers.* University of Missouri Press, Columbia, 1940. o.p.

Cutright, Paul Russell. *Lewis and Clark: Pioneering Naturalists.* University of Illinois Press, Urbana, 1969.

DeVoto, Bernard. *The Course of Empire.* Houghton Mifflin Company, Boston, 1952.

———. *The Journals of Lewis and Clark.* Houghton Mifflin Company, Boston, 1953.

Wheeler, Olin D. *The Trail of Lewis and Clark, 1804-1904.* G. P. Putnam's Sons, New York and London, 1926. o.p

PRINTED IN U.S.A.

ABOUT THE AUTHOR

Ingvard Henry Eide was once a cemetery lot salesman, the first commentator to emcee a Hollywood musical production on radio, a farm and country editor, the public relations agent for the Flathead Indians in Montana, and a friend of the late Jim Thorpe. Mr. Eide spends all the time he can in various areas of the northern Rocky Mountains. In his retracing of the actual route taken by Lewis and Clark for AMERICAN ODYSSEY, he spent over two years and traveled over 57,000 miles. He is a western history fan, an amateur gardener, and a collector of Rocky Mountain flora. He attended the Los Angeles Art Center and was a student of Ansel Adams and the late Will Connell. As a free-lancer, Mr. Eide does photographic assignments in national architectural, advertising, and industrial publications as well as for general magazines, and special photographic projects for the U. S. Forest Service.